Physics Research and Technology

nova
science publishers
www.novapublishers.com

Physics Research and Technology

Ferrite Materials and Technologies
Ravi Panwar, PhD (Editor)
Dharmendra Singh (Editor)
2023. ISBN: 979-8-89113-086-9 (Hardcover)

Fields of Particles with Spin, Theory and Applications
Alina Ivashkevich, PhD (Editor)
Nina Krylova, PhD (Editor)
Elena Ovsiyuk, PhD (Editor)
Vasiliy Kisel, PhD (Editor)
Vladimir Balan, PhD (Editor)
Viktor Red'kov, PhD (Editor)
2023. ISBN: 979-8-89113-012-8 (Hardcover)
2023. ISBN: 979-8-89113-113-2 (eBook)

75 Years of the Pion
Igor Strakovsky, PhD (Editor)
Alexey A. Petrov, PhD (Editor)
Nicolai Popov, PhD (Editor)
2023. ISBN: 979-8-88697-917-6 (Hardcover)
2023. ISBN: 979-8-89113-164-4 (eBook)

Conjugate Construction of Quantum Optics:
From Foundations to Applications
Abhishek Kanoungo, PhD
Jeffrey Zheng, PhD (Editor)
2023. ISBN: 979-8-88697-651-9 (Hardcover)
2023. ISBN: 979-8-88697-653-3 (eBook)

More information about this series can be found at
https://novapublishers.com/product-category/series/physics-research-and-technology/

Mário F. S. Ferreira
Editor

Photonic Crystal Fibers (PCFs)

Advances in Research and Applications

nova
science publishers
www.novapublishers.com

Library of Congress Cataloging-in-Publication Data

ISBN: 979-8-88697-618-2

Published by Nova Science Publishers, Inc. † New York

Contents

Preface

The phenomenon of total internal reflection, responsible for guiding of light in optical fibers, is known from the nineteenth century. In fact, the transmission of light along a curved dielectric cylinder was the subject of a lecture demonstration by John Tyndall in 1854. His light pipe was a stream of water emerging from a hole in the side of a tank which contained a bright light. Glass fibers were fabricated since the 1920s, but their use remained restricted to medical applications until the 1960s. The use of such glass fiber for optical communications was not practical, due to their high losses (~1000 dB/km).

In 1966, Kao and Hockam suggested that optical fibers could be used in communication systems if their losses were reduced below 20 dB/km. Following an intense activity on the purification of fused silica, such main objective was achieved in 1970. Further technological progress allowed the reduction of fiber loss to 0.2 dB/km near the 1550 nm spectral region by 1979. Actually, optical fibers provide low-loss transmission over an enormous frequency range of about 25THz, which is orders of magnitude more than the bandwidth available in copper cables or any other transmission medium. The superior properties of optical fibers led to a revolution in the field of lightwave communication systems.

Actually, besides serving as transmission links, optical fibers have been used also for sensing, as well as to realize many types of photonic devices, such as couplers, interferometers, amplifiers, lasers, broad-band sources, etc.

More recently, new types of fibers, called photonic crystal fibers (PCFs), have been developed and they are characterized by the fact that the silica cladding presents an array of embedded air holes. PCFs can be divided into two main types. One type has a solid core, in which the light is guided due to total internal reflection, since the core has a higher refractive index than the cladding. The other type of PCFs has a central region containing air. Such fibers are usually called hollow-core photonic crystal fibers (HC-PCFs) and the light propagating in most of them is confined to the core by the photonic bandgap or the inhibited-coupling mechanisms.

This book is intended to provide an overview of the recent advances and applications of photonic crystal fibers. Chapter 1 provides an overview of both tapered, and photonic crystal fibers, paying a particular attention to the peculiar dispersive and nonlinear properties which make them more suitable for supercontinuum generation. The optical properties of a photonic crystal fiber are shown to depend significantly on its structural parameters, namely the hair-hole diameter and the hole-to-hole spacing.

The guiding mechanisms of hollow-core photonic crystal fibers (HC-PCFs) are reviewed in Chapter 2, a special attention being paid to kagomé HC-PCFs. Such fibers offer a broadband transmission and weak anomalous dispersion from the UV to the near-IR. These properties not only help support ultrafast soliton dynamics, but also allow the guidance of any UV light that is subsequently generated. The soliton self-compression and the possibility of deep UV dispersive wave generation in gas-filled kagome HC-PCFs is discussed.

Chapter 3 discusses the guidance mechanisms and the salient properties of both conven¬tional and photonic crystal fibers, deserving a special emphasis on the nonlinear processes and applications, such as su¬percontinuum light source and gas-filled lasers. Metal-filled HC-PCFs have shown great importance in sensor applications. In¬teresting futuristic new research areas are also included, such as Dirac mode guidance in HC-PCFs, for which special materials need to be explored.

Actually, one of the most impressive nonlinear phenomena that can be observed in highly nonlinear fibers is the supercontinuum generation (SCG). Chapter 4 discusses the main physical mechanisms contributing to the SCG process in a photonic crystal fiber. Both long duration and short (femtosecond) pump pulses are considered. The role of soliton effects in SCG become particularly important when pumping with femtosecond pulses in the anomalous dispersion region of the fiber.

Controlling the dispersion profile constitutes an important task in order to achieve a highly nonlinear and near-zero flat dispersion PCF for mid-IR applications. Chapter 5 describes the realization of wide bandwidth and high gain parametric sources based on dispersion engineered highly nonlinear multi-material PCFs. Some configurations providing a fiber optic parametric amplifier with a 3dB bandwidth of more than 2000 nm in the near IR region are proposed.

In recent years, photonic crystal fibers have attracted significant interest for specific applications at terahertz frequencies. Chapter 6 discusses different designs and interesting properties that makes PCFs suitable for various applications in the THz frequency range. A special attention is paid to THz sources of narrow and broadband radiation, and polarization devices based on various polymeric materials as the host as well as the effect of micro-inclusions in the core or the cladding configurations.

One of the problems arising from high-power injection in the optical fiber is the probability of triggering the fiber fuse effect. This effect is related to a heat conduction process in optical fibers. Chapter 7 reports on the generation of a fiber fuse and its propagation in kagome-lattice HC-PCFs and revolver fibers. It is shown that the inclusion of metallic impurities in the silica struts of a kagome HC-PCF is responsible for fiber fuse initiation. The calculated velocities of fiber fuse propagation are found to be in fair agreement with the experimental values.

I hope that the publication of this book will be useful to researchers and will facilitate further developments in this fascinating area.

Mário F. S. Ferreira
Aveiro, Portugal

Chapter 1

The Modelling of Highly Nonlinear Tapered and Photonic Crystal Fibers

Mário F. S. Ferreira*

I3N-Institute of Nanostructures, Nanomodelling and Nanofabrication, Department of Physics, University of Aveiro, Aveiro, Portugal

Abstract

Several types of highly nonlinear fibers, namely tapered and photonic crystal fibers with optimized designs, have been developed during the recent years in order to enhance their nonlinear characteristics. This chapter provides an overview of such types of fibers, paying a particular attention to the peculiar dispersive and nonlinear properties which make them more suitable for supercontinuum generation. An accurate numerical modelling is performed using a finite element numerical approach. The dependence of the dispersion properties on wavelength are calculated by solving the eigenvalue equation for fiber modes and using Sellmeier equation. The optical properties of tapered fibers are shown to be very sensitive to the core size. The dependence of the dispersion properties and of the effective mode area of a photonic crystal fiber on the air-hole diameter and the hole-to-hole spacing is also reviewed.

Keywords: highly nonlinear fibers, tapered fibers, photonic crystal fibers, nonlinear fiber optics, finite element method

---

* Corresponding Author's Email: mfernando@ua.pt.

In: Photonic Crystal Fibers (PCFs)
Editor: Mário F. S. Ferreira
ISBN: 979-8-88697-618-2

1. Introduction

Conventional glass fibers for optical communications are made of fused silica and show an attenuation as low as 0.148 dB/km [1] with a broad low-loss optical window which covers about 60 THz, ranging from 1260 to 1675 μm [2]. A number of third order nonlinear processes can occur when high-power light is launched into such low-loss optical fibers [3]; these can grow to appreciable magnitudes over the long lengths available in fibers, even though the nonlinear index of the silica glass is very small ($n_2 = 2.35 \times 10^{-20}$ m^2/W at 1550 nm [4]). Among such nonlinear processes, we may refer the self-phase modulation (SPM) [5], cross-phase modulation (XPM) [6], four-wave mixing (FWM)[7], modulation instability (MI) [3] and stimulated Raman scattering (SRS) [9].

A rather interesting nonlinear phenomenon which can be observed in optical fibers is the supercontinuum generation (SCG), which corresponds to an extremely wide spectrum achieved by an optical pulse along its propagation [10]. It results generally from the synergy between some of the above mentioned nonlinear processes. The spectral locations and powers of the pumps, as well as the nonlinear and dispersive characteristics of the fiber determine the relative importance and the interaction between these nonlinear processes. A supercontinuum source can find applications in the area of biomedical optics, where it allows the improvement of longitudinal resolution in optical coherence tomography by more than an order of magnitude [11]; in optical frequency metrology [12]; in all kinds of spectroscopy, and as multiwavelength source in the telecommunications area [13].

The fiber nonlinearity is commonly characterized by the nonlinear parameter γ (defined in the next section), which in standard single-mode silica fibers has a typical value of $\gamma \approx 1.3\,\mathrm{W}^{-1}$/km [14]. Such value is too small for some applications requiring highly efficient nonlinear processes, namely for optical signal processing. As a consequence, a fiber length of several kilometres would generally be necessary in such applications. The desire to reduce such fiber length has led to the development of some types of specialised nonlinear fibers based on conventional fiber technology. Such fibers have the refractive index contrast between core and cladding increased, up to around 3%, and well controlled dispersion characteristics. These fibers are called highly nonlinear fibers (HNLFs), since they present a smaller effective mode area, and hence a larger nonlinear coefficient (

$\gamma \approx 10 - 20\ \mathrm{W^{-1}/km}$) [15]. About 1km of this type of fiber is usually used for the demonstration of different nonlinear effects, namely to achieve supercontinua spanning hundreds of nanometres based on excitation at 1500 nm wavelength.

Further improved performance has resulted from the use of microstructured optical fibres (MOFs), also called photonic crystal fibers (PCFs) [16,17]. PCFs with a small core dimension and a cladding with a large air-fill fraction allow for extremely tight mode confinement, i.e., small effective mode area, and hence, a higher value of γ. Besides, the large refractive index difference between silica and air critically cause a strong waveguide contribution to the overall group-velocity dispersion (GVD) of the guided mode [18]. As a consequence, these structures can provide anomalous group-velocity dispersion and hence conditions for soliton formation at shorter wavelengths than is possible with conventional fibres [19]. Silica nanowires can be considered as a special case of a highly nonlinear photonic crystal fibers [16,17], in which the air-filling fraction of the cladding has reached unity.

Due to the PCF heterogeneous structure, one must use full vectorial methods in order to accurately determine the electromagnetic propagation inside those fibers. Various model/numerical approaches have been proposed. Among them, there are modal decomposition approaches using sinusoidal functions [20], Hermite-Gaussian functions [21] or cylindrical functions (multiple method) [22] and more direct numerical methods, like finite difference times domain (FDTD) [23] or finite element [24] schemes. The finite element approaches have the advantage of allowing the selection of the order and the number of elements for a given required accuracy. Among the finite element methods, there are approaches only able to compute the propagating modes and others more general that can also calculate the propagation characteristics of leaky modes by using perfectly matched layers [25].

In this Chapter we provide an overview of the dispersive and nonlinear characteristics of both tapered and photonic crystal fibers. Sections 2 and 3 describe the basic aspects of optical fiber nonlinearity and dispersion, respectively. Sesion 4 is dedicated to the finite element method approach to find the properties of optical fibers. Sections 5 and 6 review the main properties of tapered and phtonic crystal fibers, respectively. Finaly, Section 6 summarizes the main conclusions.

2. Fiber Nonlinearity

Nonlinear effects are attributed to the dependence of the susceptibility on the electric field, which becomes important at high field strengths. As a result, the total polarization vector **P** can be written in the frequency domain as a power series expansion in the electric field vector [26]:

$$\mathbf{P}(\mathrm{r},\omega)=\varepsilon_0\left[\chi^{(1)}\cdot\mathbf{E}+\chi^{(2)}:\mathbf{EE}+\chi^{(3)}\vdots\mathbf{EEE}+\ldots\right]=\mathbf{P}_L(\mathrm{r},\omega)+\mathbf{P}_{NL}(\mathrm{r},\omega) \tag{1}$$

where $\chi^{(j)}$ ($j=1, 2, \ldots$) is the jth order susceptibility. To account for the light polarization effects, $\chi^{(j)}$ is a tensor of rank j+1. The linear susceptibility $\chi^{(1)}$ determines the linear part of the polarization $\mathbf{P}_L$. On the other hand, terms of second and higher order in (2.1) determine the nonlinear polarization $\mathbf{P}_{NL}$. Since SiO_2 is a symmetric molecule, the second-order susceptibility $\chi^{(2)}$ vanishes for silica glasses. As a consequence, virtually all nonlinear effects in optical fibers are determined by the third order susceptibility $\chi^{(3)}$. In time domain, the form of the expansion is identical to (1) if the nonlinear response is assumed to be instantaneous.

The presence of $\chi^{(3)}$ implies that the refractive index depends on the field intensity, I, in the form [3]

$$n=\sqrt{1+\chi^{(1)}+\frac{3}{2}\frac{\chi^{(3)}}{c\varepsilon_0 n_0}I}\approx n_0+n_2 I \tag{2}$$

where $n_0=\sqrt{1+\chi^{(1)}}$ is the linear refractive index and $n_2=3\chi^{(3)}/(4c\varepsilon_0 n_0^2)$ is the refractive index nonlinear coefficient, also known as the *Kerr coefficient*. A value of $n_2=2.73\times10^{-20}\,\mathrm{m^2/W}$ has been measured at a wavelength of 1.06 μm [27] and several other measurements indicate that this value decreases with increasing wavelengths [28]. Considering the telecommunications window, a value $n_2=2.35\times10^{-20}\,\mathrm{m}^2$

/W has been obtained [4]. In spite of this small value, the effects of the nonlinear refractive index become significant due to very long interaction lengths provided by the low loss optical fibers.

The fiber nonlinearity is commonly characterized by the nonlinear parameter γ, which is given by [3]

$$\gamma = \frac{2\pi}{\lambda} \frac{n_2}{A_{eff}} \tag{3}$$

where λ is the light wavelength, n_2 is the nonlinear-index coefficient of the fiber core and A_{eff} is the effective mode area, given by:

$$A_{eff} = \frac{\left(\iint |F(x,y)|^2 dxdy \right)^2}{\iint |F(x,y)|^4 dxdy} \tag{4}$$

$F(x, y)$ representing the spatial distribution of the fiber mode.

Eq. (3) shows that, for a fixed wavelength, since n_2 is determined by the material from which the fiber is made, the most practical way of increasing the nonlinear parameter γ is to reduce the effective mode area A_{eff}. However, the nonlinear parameter γ can also be enhanced using the dopant dependence of the nonlinear refractive index n_2. It has been shown that n_2 increases by doping the bulk glass with GeO_2 [29]. On the other hand, n_2 decreases in the case of a F-doped bulk glass.

In order to enhance the nonlinear parameter γ, the best option will be provided by a fiber heavily doped with GeO_2 in the core and having a large refractive index difference between the core and the cladding, because such fibers exhibit a large n_2 as well as a small effective area A_{eff} [30]. Optical fibers designed in this way are known as highly nonlinear fibers (HNLFs). In typical HNLFs, the relative refractive index difference of the core to the outer cladding is around 3%, while the core diameter is around 4μm. The values of the nonlinear parameter γ for most of these fibers are in the range 10 to 20

W^{-1}/km [15]. Increasing γ much above these values is not possible in these types of fibers, since the optical mode confinement is lost when the core diameter is further reduced.

3. Fiber Dispersion

The propagation modes of an optical fiber can be obtained by solving the Helmholtz equation [31]:

$$\vec{\nabla}\times(\vec{\nabla}\times\vec{E})-n^2k_0^2\vec{E}=0 \tag{5}$$

and considering the appropriate boundary conditions. In Eq. (3), n is the local refractive index, k_0 is the free space wave vector and $\vec{E}$ is the electric field, which can be written in the form:

$$\vec{E}(r,\phi,z)=\vec{E}(r,\phi)e^{-i\beta z} \tag{6}$$

$\beta(\omega)$ being the mode-propagation constant. $\beta(\omega)$ is related with the refractive index $n(\omega)$ in the form

$$\beta(\omega)=\frac{\omega n(\omega)}{c} \tag{7}$$

The refractive index of fused silica depends on the wavelength according with the Sellmeir equation [32]:

$$n^2=1+\frac{0.6961663\,\lambda^2}{\lambda^2-(0.0684043\,)^2}+\frac{0.4079426\,\lambda^2}{\lambda^2-(0.1162414\,)^2}+\frac{0.8974794\,\lambda^2}{\lambda^2-(9.896161\,)^2} \tag{8}$$

Mathematically, the effects of fiber dispersion are accounted for by expanding $\beta(\omega)$ in a Taylor series about the carrier frequency ω_0 at which the pulse spectrum is centered:

$$\beta(\omega) = \beta_0 + \beta_1(\omega - \omega_0) + \frac{1}{2}\beta_2(\omega - \omega_0)^2 + \ldots \tag{9}$$

where

$$\beta_j = \left(\frac{d^j\beta}{d\omega^j}\right)_{\omega=\omega_0} \quad (j = 0, 1, 2, \ldots) \tag{10}$$

The group velocity dispersion (GVD) is characterized by the parameter β_2

$$\beta_2 = \frac{d^2\beta}{d\omega^2} = -\frac{dv_g / d\omega}{v_g^2} \tag{11}$$

where $v_g = c / n_g$ and $n_g = n + \omega(dn / d\omega)$ represent the group velocity and the group refractive index, respectively. In practice, the GVD is often characterized by another parameter, *D,* given by

$$D = -\frac{2\pi c\beta_2}{\lambda^2} = -\frac{\lambda}{c}\frac{d^2n}{d\lambda^2}. \tag{12}$$

The parameter *D* gives the delay of arrival time in ps unit for two wavelength components separated by 1 nm over a distance of 1 km. In the case of fused silica, $D = 0$ at $\lambda = \lambda_{ZD} = 1.276$ μm, which is referred to as the silica zero-dispersion wavelength. The parameter *D* becomes negative (normal GVD) at shorter wavelengths and positive (anomalous GVD) at longer wavelengths.

The overall GVD of a guided mode in an optical fiber depends not only on the material dispersion, but also on the waveguide dispersion [3]. The last contribution becomes especially important in the case of fibers using a large difference in refractive index between the core and cladding materials. Besides affecting the dispersion characteristics, increasing such refractive index difference provides also an enhancement of the nonlinear parameter of the fiber, as discussed previously.

When the core diameter of an optical fiber is much greater that the wavelength, the guided mode effective index must approach the refractive index of the core material. When the core diameter is far less than the wavelength, the mode effective index approaches the refractive index of the cladding material. In between these two situations there is dispersion, which depends on the refractive index difference between the core and cladding. The GVD due to the waveguide is anomalous at short wavelengths and becomes normal for long wavelengths, a behaviour which is opposite to that of the material dispersion.

In standard fibers the relative index difference between the core and cladding is less than 1% and the effects of waveguide dispersion are weak. The fiber global dispersion is given by the sum of the material dispersion, D_M, and waveguide dispersion, D_w. The main effect of waveguide dispersion is to shift λ_{ZD} so that the total dispersion is zero near 1.31 μm.

It is possible to design the fiber such that the zero-dispersion wavelength λ_{ZD} is shifted to the vicinity of 1.55 μm. Such fibers are referred to as *dispersion-shifted fibers* (DSFs) It is also possible to design the fiber such that the total dispersion D is relatively small over a wide wavelength range extending from 1.3 to 1.6 μm. Such fibers are called *dispersion-flattened fibers*.

When the relative index difference between the core and cladding is further increased and the light becomes more tightly confined the waveguide dispersion can dominate the overall GVD. This is the case of the opical fibers discussed in the following two sections.

4. Finite Element Analysis

The optical properties of tapered and photonic crystal fibers can be computed applying the finite element analysis method (FEM). FEM is usually applied to boundary-value problems [33], following some basic steps, namely: discretization of domains, selection of interpolation functions, system of equations formulation, and system of equations solution.

Discretization is probably the most important step in FEM analysis due to the fact that the domain's discretization will affect the numerical accuracy of the results obtained, while also significantly affecting computation time. In this process, the entire domain is subdivided into a number of smaller domains, which are known as elements. In a one-dimensional case, a line is divided into shorter line segments. Usually, triangular and tetrahedral shapes are used to map arbitrary two- and three-dimensional domains, respectively.

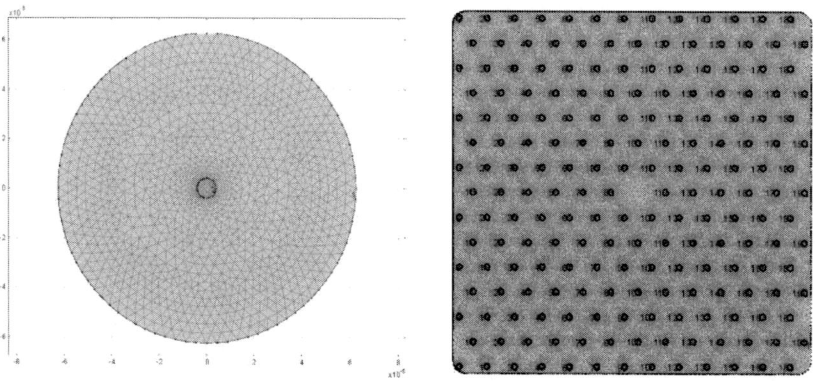

(a) (b)

Figure 1. Mesh used for (a) the tapered fiber and (b) the microstructured fiber.

When studying tapered and photonic crystal fibers, we must consider a two-dimensional domain, corresponding to a given sectional plane of these fibers. The discretization is known as creating a mesh for the problem at hand. Fig. 1a shows the triangular mesh assumed in the simulations, using the commercial software COMSOL Multiphysics [34]. Concerning the photonic crystal fiber, a hexagonal pattern was considered, as illustrated in Figure 1b.

Since FEM is only applicable to a finite or bounded domain, one needs to truncate the infinite scattering region by placing an artificial boundary, to limit the discretization domain. We must consider also a boundary condition providing a unique solution for the problem at hand. The selected condition should make the boundary as transparent as possible to the scattered field, minimizing the nonphysical reflections from it.

The second step to consider when using the FEM is to select the interpolation functions that provide an approximation of the unknown solution within an element. Interpolation functions are usually polynomial functions, mostly linear polynomials. Although higher order polynomials usually provide more accurate approximations, this normally results in a more complicated formulation, which explains the widespread use of simple and basic linear polynomial functions.

Another of the major steps in finite element analysis is formulating the system of equations. Two different methods can be used for this purpose: 1)

Ritz method and 2) Galerkin method [33]. The Ritz method is a variational method, in which the trial function is selected to meet the boundary conditions. Since the exact function is not known, such trial function contains several adjustable parameters, which are varied to find the optimal set. Concerning the Galerkin method, it is a weighted residual method, which uses an orthogonal basis to reach the solution.

The last step in the FEM is solving the system of equations. Depending on the particular type of problem at hand, we can have a deterministic-type equation or an eigenvalue-type equation to solve [33]. Regarding the field of electromagnetic waves, the last case occurs in source-free problems, including the problem of wave propagation in waveguides. In our case, the eigenvalue equation is obtained by imposing the appropriate boundary conditions to the electric and magnetic fields obeying the Helmholtz equation given by Eq. (5).

5. Tapered Fibers

A main parameter characterizing an optical fiber is the normalized frequency *V*, given by

$$V = k_0 a n_1 \sqrt{2\Delta} \tag{13}$$

where k_0 is the vacuum wavenumber, a is the core radius, n_1 is the core refractive index, and

$$\Delta = (n_1 - n_c)/n_1 \tag{14}$$

is the relative index difference between the core and the cladding of the fiber, n_c being the cladding refractive index. A single-mode fiber is achieved when $V < 2.405$. If Δ is kept constant and the core radius a is reduced, the normalized frequency V decreases and the mode confinement is lost. Eq. (11) shows that, in order to maintain a given value of V when decreasing a, Δ must increase such that $a^2\Delta$ remains constant.

The maximum value of Δ is achieved when the cladding material is replaced by air ($n_c \approx 1$), in which case the fiber becomes the so-called jacketed air-suspended rod (JASR). This kind of fibers have been produced by tapering standard optical fibers, with an original cladding diameter of 125 μm [35-37]. Tapering a fiber involves heating and stretching it to form a narrow waist connected to untapered fiber by taper transitions. If the transitions are gradual, light propagating along the fiber suffers very little loss.

Figure 2 shows the fundamental mode of a tapered silica fiber surrounded by air, for two different fiber diameters: (a) d = 2 μm and (b) d = 0.5 μm. In the case of Figure 2a, the field is completely confined by the fiber. However, in the case of Figure 2b the fiber cannot completely confine the propagating wave, leading to a relatively large evanescent field.

The nonlinear parameter γ is significantly enhanced when the core diameter is reduced to values of the order of 2 μm. Considering the Gaussian approximation for the mode profile, and using $\lambda = 1$ μm for the wavelength of light transmitted through the fiber, $n_1 = 1.45$ for the silica core and $n_c = 1$ for the air cladding, it was found that a maximum value $\gamma \approx 370$ W^{-1}/km occurs for $V \approx 1.85$ [38].

Figure 3 shows the mode spot size as a function of the fiber diameter, considering different fiber materials: silica (SiO_2), a lead silicate glass ($SiO_2 - PbO$), and a bismuth silicate glass ($SiO_2 - Bi_{12}O_{18}$). We observe the existence of a minimum beam waist for a given value of the fiber diameter, which depends on the fiber material. Two regions are identified in Figure 3. Region I corresponds to the large evanescent-field region, which may be useful for sensing applications. Region II corresponds to the high-confinement region, which is more appropriate for nonlinear applications.

In recent years there has been a great interest in tapered fibers with a core diameter below 1 μm [39-41]. Numerical results, obtained without using the Gaussian approximation, have shown that the mean-field diameter of the optical mode of such a nanoscale fiber with silica core and air cladding achieves a minimum value when the core diameter is 0.74λ [40]. The nonlinear parameter γ attains then a maximum value, which scales with λ^{-3}. Using $n_2 = 2.6 \times 10^{-20}$ m^2/W a value $\gamma = 662$ W^{-1}/km is obtained at $\lambda = 0.8$ μm.

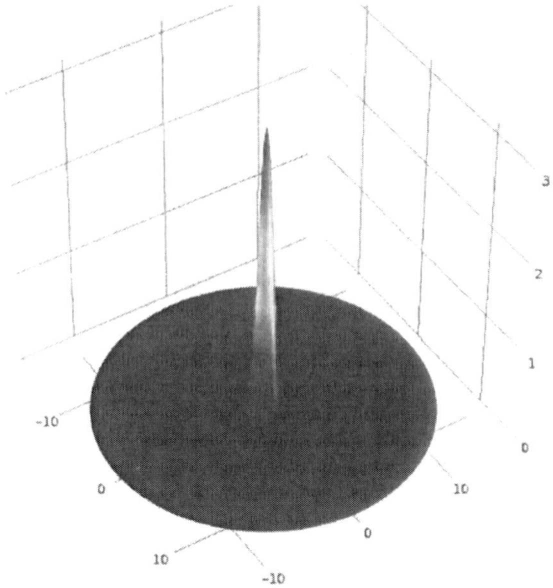

(a)

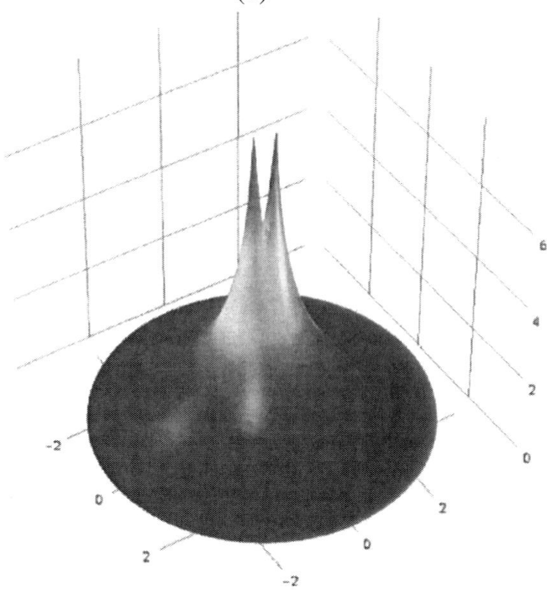

(b)

Figure 2. Fundamental mode of a tapered silica fiber surrounded by air with a diameter: (a) d = 2 μm and (b) d = 0.5 μm.

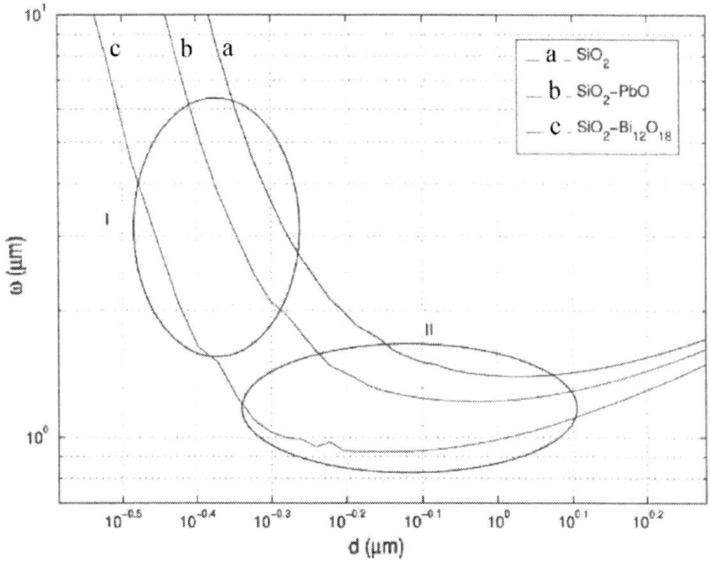

Figure 3. Variation of the spot size of a tapered fiber with the fiber diameter.

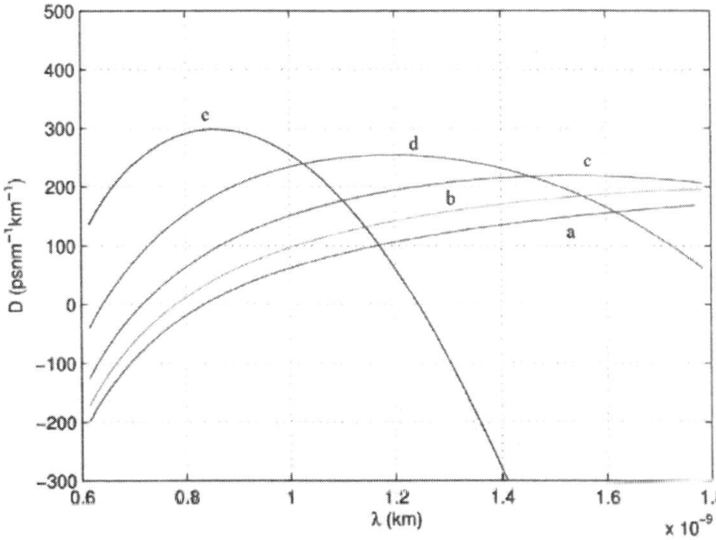

Figure 4. Dispersion properties of a tapered fiber surrounded by air, for five values of the diameter: $d = 3.0$ μm (curve a), $d = 2.5$ μm (curve b), $d = 2.0$ μm (curve c), $d = 1.5$ μm (curve d), and $d = 1.0$ μm (curve e).

The dispersive characteristics of a tapered fiber are very sensitive and can be adjusted by changing the fiber diameter [40-42]. This is illustrated in Fig. 4, which shows the calculated dispersion spectra for several values of the taper diameter. We observe that, by reducing fiber diameter, the zero-dispersion wavelength (ZDW) can be shifted to the visible range and, in some cases, a second ZDW appears at longer wavelengths, defining a wavelength window between the two ZDWs, which shows anomalous GVD.

The impact of waveguide dispersion can also be appreciated from Fig. 5, which shows the dispersion of bulk silica (curve a). and the dispersion curve for the fundamental mode of a tapered silica fiber with a diameter of 1.0 μm (curve b), and 0.8 μm (curve c). The GVD of bulk silica crosses through zero at around 1300 nm, being positive (anomalous) at longer wavelengths and negative (normal) at shorter wavelengths. We observe that the waveguide dispersion dominates clearly the overall dispersion in the tapered fiber case. In particular, the normal waveguide dispersion at long wavelengths dominates over the anomalous material dispersion, whereas both ZDWs of the tapered fiber are below the material ZDW.

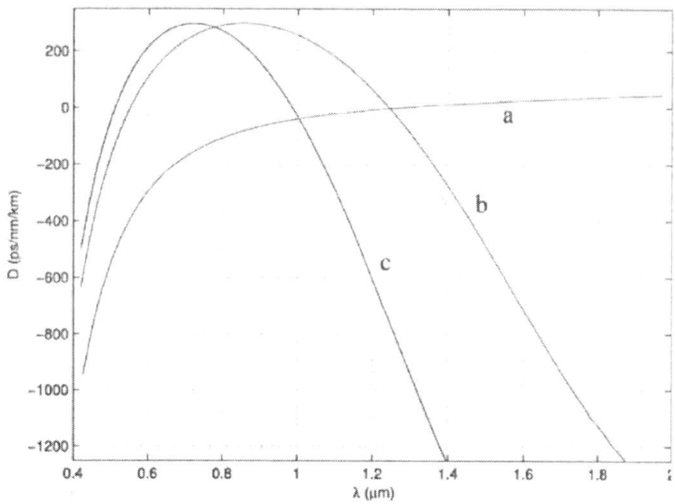

Figure 5. Dispersion curves for the case of bulk silica (curve a). and for the fundamental mode of a tapered silica fiber with a diameter of 1.0 μm (curve b), and 0.8 μm (curve c).

For efficient supercontinuum generation in an optical fiber, it is generally desirable that the dispersion of the fiber guided mode be zero or anomalous at the pump wavelength. Er^{3+}-based sources at 1550 nm are suitable for conventional step-index silica fibers. However, pump sources based around 1060 nm and 800 nm can also be used in the case of fiber with a high index difference between the core and cladding, adjusting appropriately their core diameter.

6. Photonic Crystal Fibers

Photonic crystal fibers (PCFs), also known as microstructured optical fibers (MOFs), represent a new class of optical fibers that are characterized by the fact that the silica cladding presents an array of embedded air holes. They were first realized in 1996 in the form of a photonic crystal cladding with a periodic array of air holes [16].

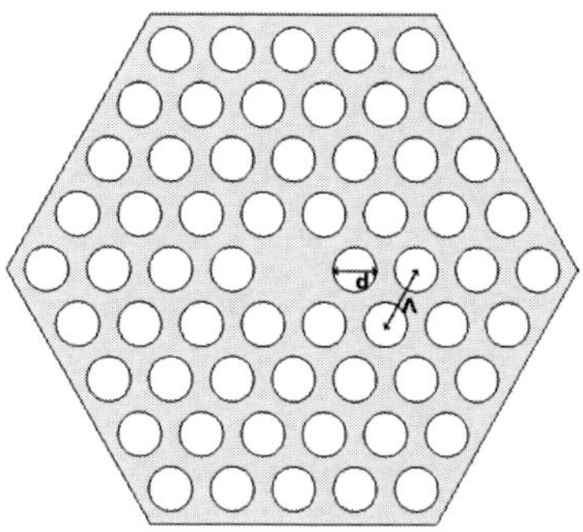

Figure 6. Schematic of a solid-core photonic crystal fiber with a hexagonal geometry. The two main parameters are the hair-hole diameter *d* and the hole-to-hole spacing Λ.

Photonic crystal fibers can be divided in two main types. One class of fibers, first proposed in 1999 [43], have a central region containing air. Such fibers are usually called hollow-core PCFs and the light propagating in them is confined to the core by the photonic bandgap effect. The nonlinear effects are strongly reduced and the dispersion becomes negligible in this kind of PCFs. A nonlinear coefficient $\gamma = 0.023\ \mathrm{W^{-1}/km}$ was reported recently for a hollow core fiber [44]. However, the nonlinear effects in this type of PCFs can be greatly enhanced if air is replaced with a suitable gas or liquid [45].

The second type of photonic crystal fibers has a solid core, in which the light is guided due to total internal reflection, since the core has a higher refractive index than the cladding. A solid core PCF with an hexagonal pattern of holes is represented in Fig. 6; it has two main structural parameters: the air-hole diameter d and the hole-to-hole spacing Λ. In such fibers the periodic nature of air holes in the cladding is not important as long as they provide an effective reduction of its refractive index below that of the silica core [46, 47]. This helps to concentrate the mode field in a very small area, which is particularly the case in PCFs with and large air-filling fractions. The tapered fiber considered above can be used to model the behavior of the fundamental mode of air-silica photonic crystal fibers with a high air-filling fraction in their cladding [48].

The number of guided modes in a conventional fiber is determined by the normalized frequency V given by Eq. (13), which depends on the wavelength. In conventional fibers, the single-mode condition $V < 2.405$ is only satisfied if the wavelength is in the infrared region. For lower values of the wavelength the fiber becomes multimode. However, in a MOF the normalized frequency is given by a slightly different expression from Eq. (13), with a replaced by Λ and n_c replaced by the effective cladding index of the microstructured cladding. Such effective refractive index depends significantly on the ratios d/Λ and λ/Λ [49] and asymptotically approaches the core index in the short wavelength limit. As a consequence, both the effective index difference and the numerical aperture go to zero in this limit.

Surprisingly, in a MOF it is verified that the behavior of the cladding effective index cancels out the wavelength dependence of the normalized frequency V, such that it approaches a constant value as $\Lambda/\lambda \to \infty$. As a consequence, it was observed that a MOF shows a single mode behaviour at all wavelengths if $d/\Lambda < 0.43$; such fiber is known as the *endlessly single-mode* (ESM) fiber [50,51]. Furthermore, because of the scale invariance of MFs, the ESM behavior can be achieved for all core diameters.

Figure 7 shows the spatial mode distribution for a photonic crystal fiber with a hexagonal pattern of holes spaced by $\Lambda = 10\ \mu\text{m}$ and of diameter (a) d = 0.5 μm, (b) d = 2 μm,, and (c) d = 4 μm, considering a wavelength $\lambda = 1.55\ \mu\text{m}$. We observe a clear reduction of the mode area when the hole diameter is increased, which determines an increased air-filling fraction. Figure 8 displays the normalized effective mode area against the normalized wavelength for different hole-pitch ratios. We confirm that increasing the hole-pitch ratio the effective mode area is reduced, which becomes more significant for lower values of the wavelength. Consequently, the nonlinear parameter is increased, as given by Eq. (1).

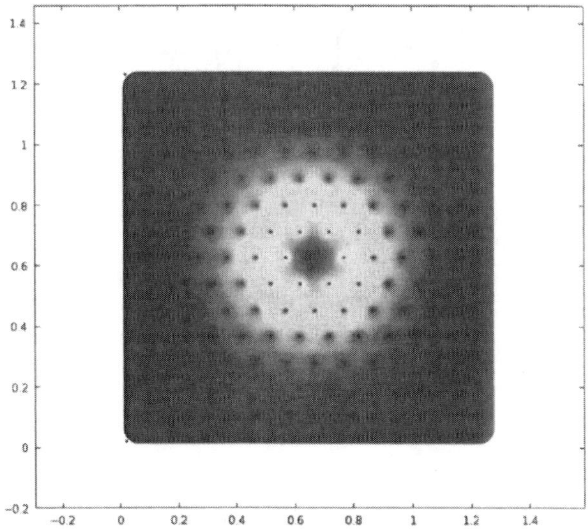

(a)

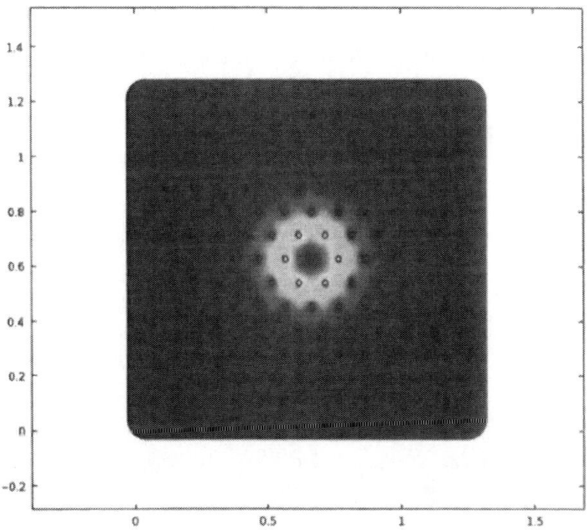

(b)

Figure 7. (Continued).

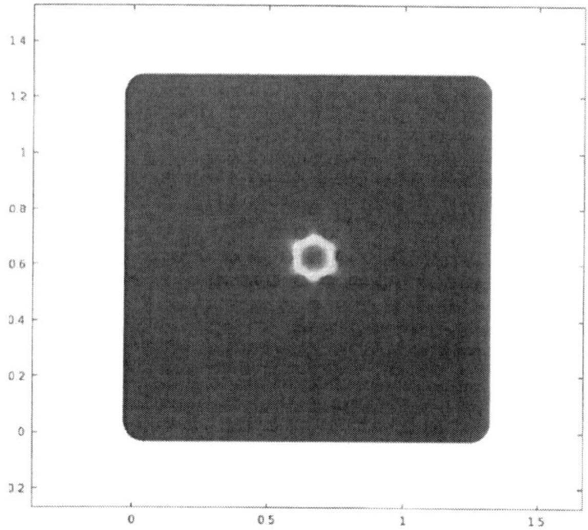

(c)

Figure 7. Spatial mode distribution for a photonic crystal fiber with a hexagonal pattern of holes spaced by $\Lambda = 10$ μm and of diameter (a) $d = 0.5$ μm, (b) $d = 2$ μm,, and (c) $d = 4$ μm, considering a wavelength $\lambda = 1.55$ μm.

In the case of photonic crystal fibers and to take into account the different proportions of light in glass and air, the nonlinear coefficient γ must be redefined as follows [52]:

$$\gamma = k_0 \sum_i \frac{n_2^i}{A_{eff}^i} = k_0 \frac{n_2^{eff}}{A_{core}} \tag{15}$$

In Eq. (15) n_2^i is the nonlinear refractive index of material i (2.9×10^{-23} m^{2}W^{-1} for air, 2.6×10^{-20} m^{2}W^{-1} for silica), A_{eff}^i is the effective area for the light in material *i*, and n_2^{eff} is the effective nonlinear index for fiber, with a core area A_{core}. A nonlinear coefficient $\gamma = 240$ W^{-1}/km at 850 nm was measured for a solid-core PCF with a core diameter of 1μm [15].

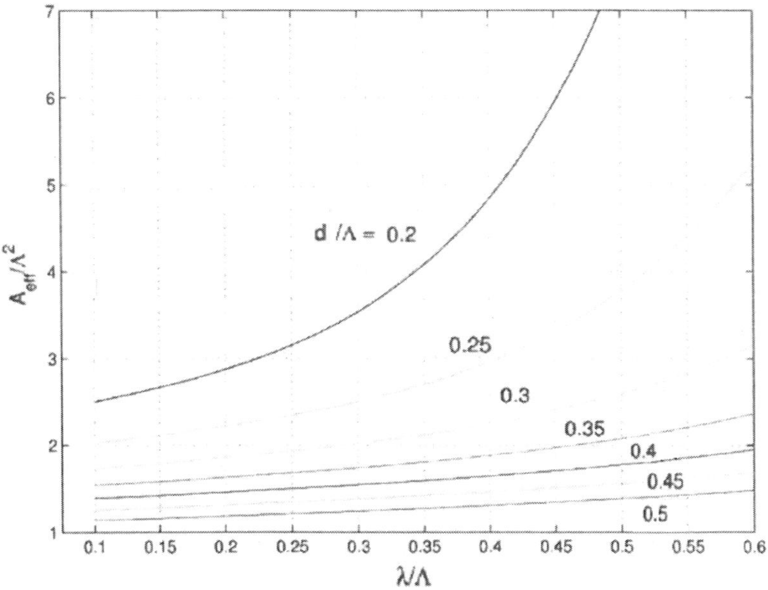

Figure 8. Normalized effective mode area against the normalized wavelength for different hole-pitch ratios.

A peak power of $P_0 = 5$ kW launched into a silica PCF with a nonlinear coefficient $\gamma = 240\ \text{W}^{-1}/\text{km}$ provides a nonlinear length $L_{NL} = (\gamma P_0)^{-1} < 1$ mm. For dispersion values in the range $-400 < \beta_2 < 400\,\text{ps}^2/\text{km}$ and pulse durations $t_0 = 300$ fs, we have a dispersion length $L_D = t_0^2 / |\beta_2| > 0.2$ m. Considering also typical values of fiber loss (α ~1-100dB/km), we conclude that both the effective fiber length $L_{eff} = (1 - \exp(-\alpha L))/\alpha$ and the dispersion length L_D are much longer than the nonlinear length L_{NL}, which means that strong nonlinear effects will be readily observable in solid-core PCFs.

The microstructured cladding offers greatly enhanced design flexibility and can manipulate the dispersion characteristics by controlling structural parameters such as the hair-hole diameter *d* and the hole-to-hole spacing Λ [49]. In fact, the dispersive properties of PCFs are quite sensitive to these parameters and can be tailored by changing appropriately each of them. Figure 7 shows the dispersion characteristics of a photonic crystal fiber with a hexagonal pattern of holes spaced by (a) $\Lambda = 2\,\mu\text{m}$ and (b) $\Lambda = 1\,\mu\text{m}$,, for different hole-pitch ratios.

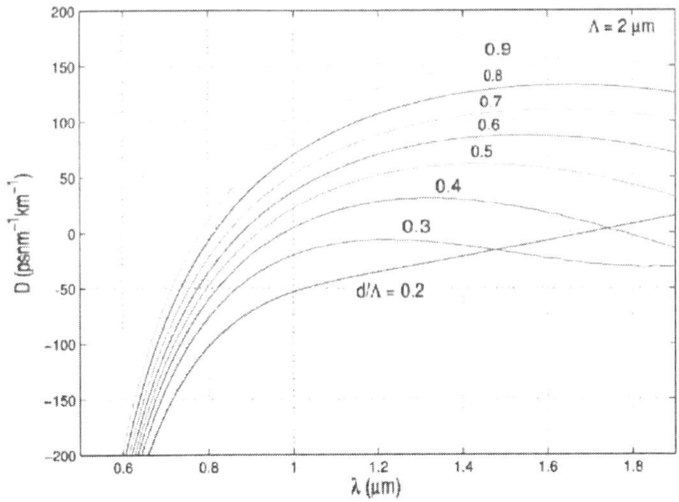

(a)

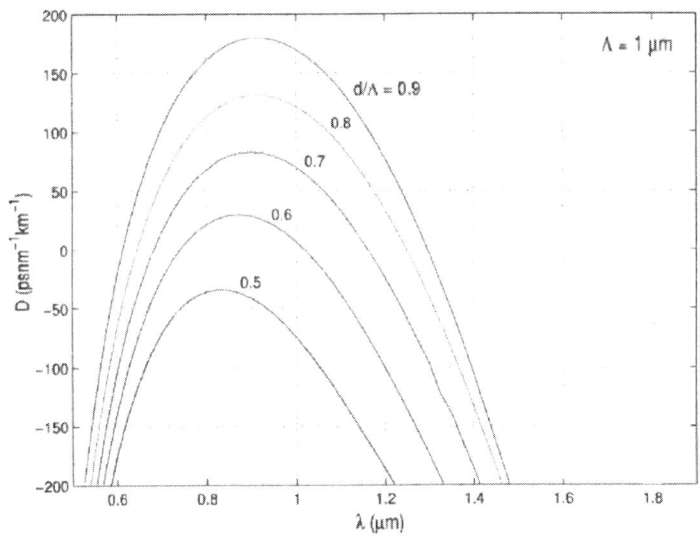

(b)

Figure 9. Dispersion properties of a photonic crystal fiber with a hexagonal pattern of holes spaced by by (a) $\Lambda = 2\,\mu\text{m}$ and (b) $\Lambda = 1\,\mu\text{m}$,, for different hole-pitch ratios.

In solid-core PCF, as the holes get larger, the core becomes more and more isolated, until it resembles an isolated strand of silica glass. The core diameter of a PCF is given by $d_{core} = 2\Lambda - d$. PCFs with larger cores exhibit semi-infinite anomalous dispersion above the ZDW. By decreasing the core size, the ZDW tends to be shifted to a shorter wavelength, leading to the anomalous dispersion at near infrared and visible wavelengths. When the core size is decreased further, a second ZDW arises in the longer wavelength side, such that the GVD is anomalous in the spectral window between the two ZDWs and normal outside it. This situation can be seen in some cases of Fig.s 9(a) and (b). Submicron-diameter PCF cores have been fabricated using a conventional tapering process [42].

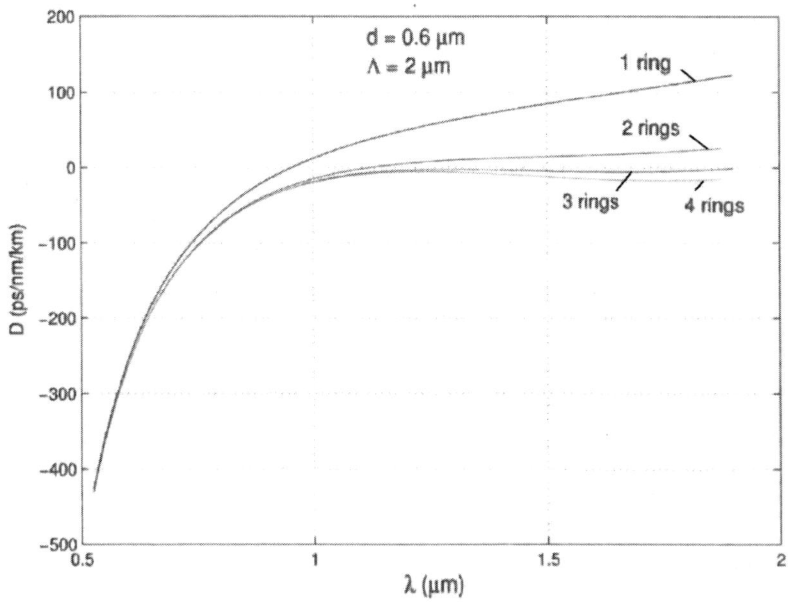

Figure 10. Dispersion properties of a photonic crystal fiber with a hexagonal pattern of holes spaced by $\Lambda = 2$ μm and hole diameter d = 0.6 μm at $\lambda = 1.55$ μm, for different number of hole rings around the fiber core.

A nearly zero dispersion and flattened behaviour is observed in some cases of Fig. 9. This characteristic can be further controlled by adjusting the number of hole rings around the core of a PCF, as illustrated in Fig. 10 for the case $\Lambda = 2$ μm and d = 0.6 μm. The availability of fibers with nearly zero dispersion is particularly important for supercontinuum generation and other telecom applications making use of the four-wave mixing effect.

The remarkable features of photonic crystal fibers provides a plethora of nonlinear mechanisms of spectral broadening that may differ depending on waveguide dispersion profile, power, wavelength and duration of pumping pulses. Using such fibers, it becomes possible to generate a supercontinuum extending from the visible to the near infra-red region. Supercontinumm generation has been achieved in different PCFs at 532 [42], 647 [53], 1064 [54], and 1550 nm [55].

The dispersive properties of the fiber become especially important in the case of the FWM process. The phase-matching condition for the degenerate FWM process requires that dispersive effects compensate nonlinear ones through the following equation [3]:

$$\beta_2 \Omega_s^2 + \frac{\beta_4}{12}\Omega_s^4 + 2\gamma P_p = 0 \tag{16}$$

where β_2 and β_4 are, respectively, the second- and fourth-order dispersion terms at the pump frequency, Ω_s is the frequency shift between the signal or idler frequencies and the pump frequency, γ is the nonlinear parameter, and P_p is the pump peak power. The condition (16) can be satisfied in the case of a highly nonlinear fiber by adjusting appropriately its dispersive characteristics. In the anomalous dispersion regime ($\beta_2 < 0$), the β_2 term dominates in Eq. (16). However, the FWM effect can also occur in the normal-GVD regime ($\beta_2 > 0$). In this case, the positive nonlinear phase mismatch $2\gamma P$ is compensated by the negative value of the linear phase mismatch due to β_4. This condition can be easily realized considering the dispersive characteristics of both tapered and in photonic crystal fibers [54, 56].

Conclusion

In this chapter we presented an overview of the main nonlinear and dispersive characteristics of some types of highly nonlinear fibers, namely tapered and photonic crystal fibers. An accurate numerical modelling was performed using a finite element numerical approach. The dependence of the dispersion properties on wavelength was calculated by solving the eigenvalue equation for fiber modes and using Sellmeier equation. The optical properties of tapered fibers were shown to be very sensitive to the core size. The dependence of the dispersion properties and of the effective mode area of a photonic crystal fiber on the air-hole diameter and the hole-to-hole spacing was also reviewed. The peculiar dispersive characteristics of both tapered and photonic crystal fibers make them particularly suitable to generate the supercontinuum using commercially available pump sources around 1060 nm and 800 nm.

Founding

Fundação para a Ciência e a Tecnologia (FCT) (Project UID/CTM/ 50025/2022).

References

[1] Chigusa, Y., Yamamoto, Y., Yokokawa, T., Sasaki, T. Taru. T., Hirano, M., Kakui, M., Onishi, M. & Sasaoka, E. (2005). Low-loss pure-silica-core fibers and their possible impact on transmission systems. *J. Lightwave Technol.*, *23*, 3541-3550.

[2] Optical System Design and Engineering Consideration, *ITU-T Recomendation Series G Supplement*, *39*, Feb 2006.

[3] Ferreira, M. (2011). *Nonlinear Effects in Optical Fibers*; John Wiley & Sons, Hoboken, New Jersey.

[4] Broderick, N. G., Monroe, T. M., Bennett, P. J., Richardson, D. J. (1999). Nonlinearity in holey optical fibers: measurement and future opportunities, *Opt. Lett.*, *24*, 1395-1397.

[5] Stolen, R. H., Lin, C. (1978). Self-phase modulation in silica optical fibers., Rev.A, *17*, 1448-1453.

[6] Chraplyvy, A. R. & Stone, J. (1984). Measurement of crossphase modulation in coherent wavelength-division multiplexing using injection lasers. *Electron.Electron. Lett.*, *20*, 996-997.

[7] Stolen, R. H., Bjorkhholm, J. E. & Ashkin, A. (1974). Appl. Phys. Lett. Measurement of crossphase modulation in coherent wavelength-division multiplexing using injection lasers. *24*, 308-310

[8] Ippen, E. P. & Stolen, R. H. (1972). Stimulated Brillouin scattering in optical fibers. *Appl. Phys. Lett.*, *21*, 539-541.

[9] Stolen, R. H., Ippen, E. P. & Tynes, A. R. (1972). Raman oscillation in glass optical waveguide.*Appl. Phys. Lett.*, *20*, 62-64.

[10] Dudley, J. & Taylor, J. (Eds). (2010). Supercontinuum Generation in Optical Fibers. Cambridge University Press, Cambridge, UK.

[11] Humbert, G., Wadsworth, W., Leon-Saval, S., Knight, J., Birks, T., Russell, P., Lederer, M., Kopf, D., Wiesauer, K., Breuer, E. & Stifter, D. (2006). Supercontinuum generation system for optical coherence tomography based on tapered photonic crystal fibre. *Opt. Express*, *14*, 1596-1603.

[12] Bellini M. & Hänsch, T. (2000). Phase-locked white-light continuum pulses: toward a universal optical frequency-comb synthesizer. *Opt. Lett.*, *25*, 1049-1051.

[13] Takara, H., Ohara, T., Yamamoto, T., Masuda, H., Abe, M., Takahashi, H. & Morioka, T. (2005). *Electron. Lett.*, *41*, 270-271.

[14] Kato, T., Suetsugu, Y., Takagi, M., Sasaoka, E. & Nishimura, M. (19995). Measurement of the nonlinear refractive index in optical fiber by the cross-phase modulation method with depolarized pump light. *Opt. Lett.*, *20*, 988-990.

[15] Hirano, M., Nakanishi, T., Okuno, T. and Onishi, M. (2009). Silica-based highly nonlinear fibers and their application. *J. Sel. Topics Quantum Electron.*, *15*, 103-113.

[16] Knight, J. C., Birks, T. A., Russell, P. St. J. & Atkin, D. M. (1996). All-silica single-mode optical fiber with photonic crystal cladding. *Opt. Lett.*, *21*, 1547–1549.

[17] Russel, P. St J. (2006). Photonic-Crystal fibers. *J. Lightwave Technol.*, *24*, 4729-4749

[18] Mogilevtsev, D., Birks, T. A. & Russell, P. St. J. (1998). Group-velocity dispersion in photonic crystal fibers. *Opt. Lett.*, *23*, 1662–1664.

[19] Gander, M. J., McBride, R., Jones, J. D. C., Mogilevtsev, D., Birks, T. A., Knight, J. C. & Russell, P. St. J. (1999). Experimental measurement of group velocity dispersion in photonic crystal fibres. *Electron. Lett. 35*, 63–34.

[20] Ferrando, A., Silvestre, E., JMiret, J., Andrés, P., and Andrés, M., (1999). Full ector analysis of a realistic photonic crystal fiber. *Opt. Lett*. 24, 276-278.

[21] Mogilevtsev, D., Birks,,T., and Russel, P. (1999). Localized Function Method for Modeling Defect Modes in 2-D Photonic Crystals. *J. Lightwave Technol.* 17, 2078.

[22] Steel, M., White, T., CMartijn de Sterke, C., McPhedran, R., and Botten, L. (2001). Symmetry and degeneracy in microstructured optical fibers
Opt. Lett., 26, 488-490.

[23] Brechet, F., Marcou, J., Pagnoux, D., and Roy, P. (2000). Complete Analysis of the Characteristics of Propagation into Photonic Crystal Fibers, by the Finite Element Method. *Opt. Fiber Technology*, 6, 181-191.

[24] Lizier, j. t., and Town, G. E. (2001) Splice losses in holey optical fibers. *IEEE Photon. Technol. Lett*. 13, 794-796

[25] Saito, K., and Koshiba, M. (2002). Full-vectorial imaginary-distance beam propagation method based on a finite element scheme: application to photonic crystal fibers. *IEEE J. Quantum Electron*. 38, 927-933

[26] Butcher, P. N., and D. N. Cotter, D. N. (1990). *The Elements of Nonlinear Optics,* Cambridge University Press, Cambridge, UK

[27] Milam, D. & Weber, M. J. (1976). Measurement of nonlinear refractive-index coefficients using time-resolved interferometry: Application to optical materials for high-power neodymium lasers, *J. Appl. Phys.*, *47*, 2497.

[28] Milam, D. (1998). Review and assessment of measured values of the nonlinear refractive-index coefficient of fused silica, *Appl. Opt.*, *37*, 546.

[29] Nakajima, K. and Ohashi, M. (2002). Dopant dependence of effective nonlinear refractive index in GeO2 – and F-doped core single-mode fibers. *IEEE Photon. Technol. Lett.*, *14*, 492-494.

[30] Okuno, T., Onishi, M., Kashiwada, T., Ishikawa, S. & Nishimura, M. (1995). Silica-based functional fibers with enhanced nonlinearity and their applications. *IEEE J. Sel. Topics Quantum Electron.*, *5*, 1385-1391.

[31] Joannopoulos, J.D., Johnson, S.G., Winn, J.N., and Meade, R.D. (2008). Photonic crystals: molding the flow of light. Princeton University Press, 2nd edition

[32] Malitson, I. H. (1965). Interspecimen comparison of the refractive index of fused silica. *J. Opt. Soc. Am.* *55*, 1205–1209.

[33] Jin, J. (1993). "*The finite Element Method in Electromagnetics*", John Wiley & Sons, Inc

[34] COMSOL AB, COMSOL Multiphysics Guide, v4.1 ed. (2010).

[35] Harbold, J., Ilday, F., Wise, F., Birks, T., Wadsworth, W. & Chen, Z. (2002). Long-wavelength continuum generation about the second dispersion zero of a tapered fiber. *Opt. Lett.*, *27*, 1558-1560.

[36] Lu, F. & Knox, W. (2004). Generation of a broadband continuum with high specral coherence in tapered single-mode optical fibers. *Opt. Express*, *12*, 347-353.

[37] Foster, M. & Gaeta, A. (2004). Opt. Ultra-low threshold supercontinuum generation in sub-wavelength waveguides. *Opt. Express*, *12*, 3137-3143.

[38] Zheltitikov, A. (2005). Gaussian-mode analysis of waveguide-enhanced Kerr-type nonlinearity of optical fibers and photonic wires. *J. Opt. Soc. Am. B* 22, 1100-1104.

[39] Tong, L.; Lou, J.; and Mazur, E. (2004). Single-mode guiding properties of subwavelength-diameter silica and silicon wire waveguides. Opt. Express, 12, 1025-1035.

[40] Foster, M.; Moll, K.; and Gaeta, A. (2004). Optimal waveguide dimensions for nonlinear interactions. Opt. Express . 12, 2880-2887.

[41] Cordeiro, C., Wadsworth, W., Birks, T. & Russel, P. (2005). Engineering the dispersion of tapered fibers for supercontinuum generation with a 1064 nm pump laser. *Opt. Lett.*, *30*, 1980-1982.

[42] Leon-Saval, S., Birks, T., Wadsworth, W., Russel, P. & Mason, M. (2004). Supercontinuum generation in submicron fibre waveguides. *Opt. Express*, *12*, 2864-2869

[43] Cregan, R., Mangan, B. & Knight, J. (1999). Single-Mode Photonic Band Gap Guidance of Light in Air. *Science*, *285*, 1537-1539.

[44] Luan, F., Knight, J., Russel, P., Campbell, S., Xiao, D., Reid, D., Mangan, B., Wiliams, D. & Roberts, P. (2004). Femtosecond soliton pulse delivery at 800 nm wavelength in hollow-core photonic bandgap fibers. *Opt. Express*, *12*, 835-840.

[45] Russell, P. (2003). Photonic crystal fibers. *Science*, *299*, 358-362

[46] Broeng, J., Mogilevstev, D., Barkou, S. & Bjarklev, A. (1999). Photonic crystal fibers: a new class of optical waveguides. *Opt. Fiber Technol.*, *5*, 305-330.

[47] Eggleton, B., Kerbage, C., Westbrook, P., Windeler, R. & Hale, A. (2001). Microstructured optical fiber devices. *Opt. Express*, *9*, 698-713.

[48] Knight, J. C., Arriaga, J., Birks, T. A., Ortigosa-Blanch, A., Wadsworth, W. J. & Russell, P. St. J. (2000). Anomalous dispersion in photonic crystal fiber. *IEEE Photon. Technol. Lett. 12*, 807–809.

[49] Saito K.; Koshiba, M. (2005). Numerical modeling of photonic crystal fibers. J. Lightwave Technol. 23, 3580-3590

[50] Birks, T.; Knight, J.; Russel, P. Endlessly (1997). Single-mode photonic crystal fiber. Opt. Lett. 22, 961-963.

[51] Dong, L.; McKay, H.; L. Fu, L. (2008). All-glass endless single-mode photonic crystal fibers. Opt. Lett. 33, 2440-2442

[52] Laegsgaard, J., Mortenson, N., Riishede, J. & Bjarklev, A. (2003). Material effects in air-guiding photonic bandgap fibers. *J. Opt. Soc. Am.*, B *20*, 2046-2051.

[53] Coen, S., Chau, A., Leonhardt, R., Harvey, J., Knight, J., Wadsworth, W. & Russel, P. (2002). Supercontinuum generation by stimulated Raman scattering and parametric four-wave mixing in photonic crystal fibers. *J. Opt. Soc. Am. B*, *19*, 753-764.

[54] Wadsworth, W., Joly, N., Knight, J., Birks, T., Biancalana, F. & Russell, P. (2004). Supercontinuum and four-wave mixing with Q-switched pulses in endlessly single-mode photonic crystal fibres. *Opt. Express*, *12*, 299-309.

[55] Ravi, V., Kumar, K., George, A., Reeves, W., Knight, J., Russell, P., Omenetto, F. & Taylor, A. (2002). Extruded soft glass photonic crystal fiber for ultrabroad supercontinuum generation. *Opt. Express*, *10*, 1520-1525.

[56] Wong, G., Chen, A., Murdoch, S., Leonhardt, R., Harvey, J., Joly, N., Knight, J., Wadsworth, W. & Russel, P. (2005). Continuous-wave tunable optical parametric generation in a photonic-crystal fiber. *J. Opt. Soc. Am. B*, *22*, 2505-2511

Chapter 2

Hollow-Core Photonic Crystal Fibers

Mário F. S. Ferreira
I3N-Institute of Nanostructures, Nanomodelling and Nanofabrication, Department of Physics, University of Aveiro, Aveiro, Portugal

Abstract

The guiding mechanisms of both photonic bandgap and inhibited coupling hollow-core photonic crystal ibers (HC-PCFs) are discussed., a special attention being paid to kagomé HC-PCFs. The ultrafast dynamics of femtosecond pulse propagation in a gas-filled kagome HC-PCFs is reviewed. By varying the gas pressure, the dispersion and nonlinear characteristics of these kind of fibers can be adjusted. A generalized Schrödinger equation is used to describe ultrashort pulse propagation in a gas-filled kagome PCF. The soliton self-compression effect is observed, providing pulses with some few fs. Such extreme self-compression can lead to a highly efficient deep-UV dispersive wave generation. An ultrafast, coherent, and tunable UV light source can be developed in this way.

Keywords: hollow-core photonic crystal fibers, photonic bandgap fibers, inhibited coupling hollow-core fibers, kagomé photonic crystal fibers, pulse compression, UV light

In: Photonic Crystal Fibers (PCFs)
Editor: Mário F. S. Ferreira
ISBN: 979-8-88697-618-2

1. Introduction

The possibility of guiding light in air has has been made possible in 1999 with the first fabrication of a hollow-core photonic crystal fiber (HC-PCF) [1]. Since then, HC-PCFs have been intensively investigated since they allow for accessing previously inaccessible fields for fiber optics or to substantially improve device performance within areas such as mid-IR gas lasers [2], broadband light sources [3], nonlinear optical effects [4], high-power pulse delivery [5], gas and liquid analytics [6], and pulse compression [7].

Light propagating in many HC-PCFs is confined to the core by the photonic bandgap effect. An example is provided by HC-PCFs with a hexagonal arrangement of holes. These type of HC-PCFs offer an effective environment for nonlinear optics in gases, providing long interaction lengths while avoiding beam diffraction [8]. The main drawback of these fibers is their intrinsically narrow transmission bandwidth determined by the bandgaps, which excludes its implementation in a large number of applications in ultrafast nonlinear optics requiring broadband guidance or guidance in the visible and UV.

An alternative HC-PCF design replaces the hexagonal lattice cladding with a kagomé lattice [9, 10]. In kagomé-type HC-PCFs the field overlap with the surrounding silica structure is particularly low [9]. In contrast to the photonic bandgap-type fibers, the main guiding mechanism in this case is based on the inhibited coupling between the core and cladding modes [9] and not the bandgap-effect.

Kagomé HC-PCFs offer a broadband transmission and weak anomalous dispersion from the UV to the near-IR. These properties not only help support ultrafast soliton dynamics, but also allow the guidance of any UV light that is subsequently generated at a relatively low loss of ~3 dB/m [11]. The possibility of adjusting the gas species and gas pressure inside the fiber core also offers a new degree of freedom over conventional fibers, providing a perfect environment for demonstrating many different nonlinear effects [12]. Using noble gases we avoid also the Raman effects, allowing one to study Kerr-related phenomena in the absence of perturbations such as the soliton self-frequency shift or Raman-induced noise.

In this chapter we describe some main properties of hollow-core photonic crystal fibers, deserving a special attention to kagomé HC-PCFs. In Section 2 we discuss the guiding properties of both photonic bandgap (PBG) and inhibited coupling (IC) HC PCFs. Section 3 is dedicated to the dispersive and nonlinear properties of gas-filled kagomé HC-PCFs, whereas in Section 4 we describe some phenomena concerning the ultrafast dynamics of femtosecond pulse propagation in such fibers. The soliton self-compression effect and the generation of dispersive wave radiation in the UV region are discussed. Section 5 summarizes the main conclusions.

2. Optical Guidance Mechanisms

Hollow-core photonic crystal fibers (HC-PCFs) can guide light inside the hollow core based on two different mechanisms. The first one relies on choosing a core index and geometry so at least some of its supported mode lie in the cladding photonic bandgap range [13]. Consequently, the core-mode cannot leak out because there is no cladding mode to couple to. The second one relies on engineering a core and a cladding in order to have a strong transverse phase mismatch between the core and cladding modes, as well as a very weak spatial overlap between them. In these circumstances, the core mode remains in the core because its coupling to the cladding modes is inhibited [14].

2.1. Photonic Bandgap Formation

The formation of photonic bandgaps (PBGs) can be explained based on the anti-resonant interactions in arrays of closely spaced resonators (also referred to as rods) in the fibre cladding [15, 16]. HC-PBGFs guide light in a central hollow core by virtue of the presence of a photonic bandgap (PBG) in the periodic surrounding holey dielectric cladding region. The PBG prevents light at well-prescribed wavelengths and angles of incidence from propagating into the cladding and hence allows tight optical confinement in a suitably designed low refractive index central defect.

Several numerical techniques have been used to describe the formation of photonic band gaps in periodic structures, namely using plane-wave expansion [17], finite element [18], multipole [19] or finite difference methods [20]. More recently, simple models were developed to explain PBG formation and guidance as a result of anti-resonant interactions in arrays of closely spaced resonators in the fibre cladding [15, 16]. In particular, the tight binding model (TBM) first proposed by Benabid et al., in 2007 [21], has been successfully applied to descriibe HC-PCFs.

In solid-state physics, the bands in the energy diagram of a crystal result from the superposition of electronic state wavefunctions of the constitutive isolated atoms. In the case of a 1D crystal with a large number, *N*, of well separated atoms, the energy states is reduced to discrete states of the isolated atom, each being *N*-degenerate. As the separation between atoms decreases, the wavefunctions of their energy states begin overlapping and the N-degenerate energy state splits by virtue of the exclusion principle. This creates a band of N-distinct extended states, the width of which increases by decreasing such separation. The net result is an energy structure exhibiting allowed bands separated by gaps.

The above picture can be applied to guided photonics wherethe electronic state energy is replaced by the effective index of the guided mode. Let us consider an array of closely spaced rods. For the case of a single rod, we can retrieve the dispersion of the commonly known fiber modes. However, for a large number of equally spaced rods, the dispersion curve shows a band structure when the separation between such rods becomes sufficiently short. The formation of these bands in an extended physical photonic structure, implies the possibility of light PBG guidance in a defect within the structure if its index and geometry is judiciously chosen.

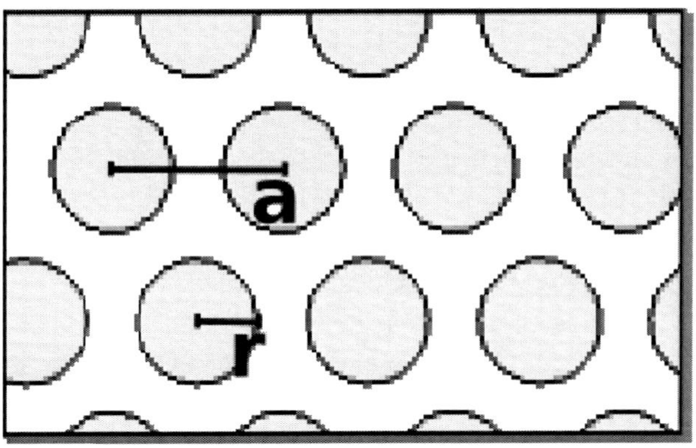

Figure 1. Scheme of a 2D photonic crystal with a rod radius r and a lattice period a.

Figure 1 illustrates a 2D photonic crystal with rod radius, r, and lattice period a,. Figure 2(a) and (b) show the corresponding transverse magnetic (TM) and transverse electric (TE) bands, respectively, for the case r/a = 0.2, considering a rod refractive index n = 3.46. The y-axis in Figure 2 represents $f/(c/a)$, where f is the light frequency, c is the vacuum speed of light, and a is the periodicity of the photonic crystal. In Figure 2(a), gaps between y = 0.275

and y = 0.446, as well as between y = 0.564 and y = 0593 are noticeable. On the other hand, in Figure 2(b) a small gap between y = 0.822 and y = 0.864 is observed.

Figure 3(a) shows a scanning electron microscopy (SEM) picture of a HC-PCF with hexagonal lattice, model HC-1550-04 from NKT Photonics. This fiber has a lattice similar to that of Figure 1. The HC-PCF has a hole diameter of 3.9μm and a pitch of 4.0μm. Figure 3(b) shows the fundamental propagation mode of this fibre at 1550nm for a core filled with air at 1atm, obtained using the COMSOL Multiphysics software.

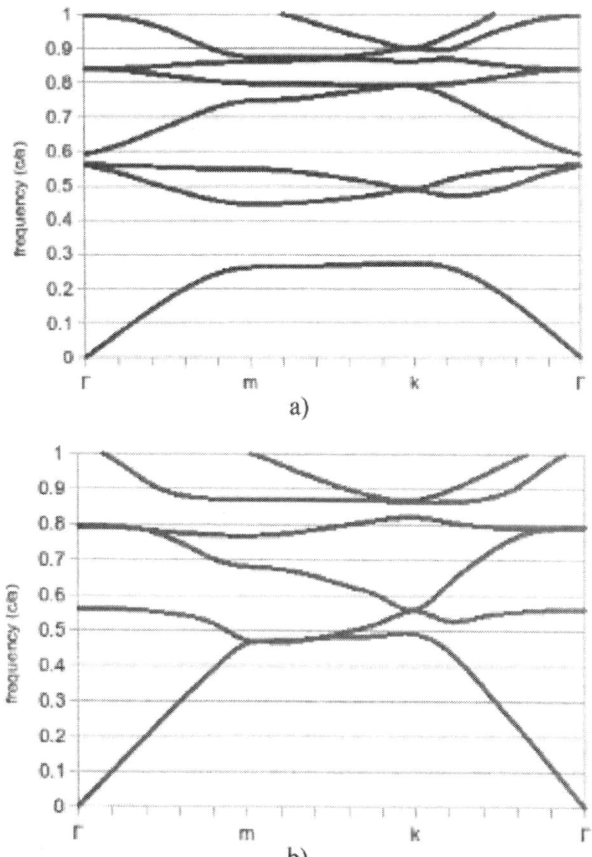

Figure 2. a) Transverse magnetic (TM) bands and b) transverse electric (TE) bands for the 2D photonic crystal represented in Figure 1 for r/a = 0.2. The rods are assumed to be surrounded by air, having a refractive index n = 3.46.

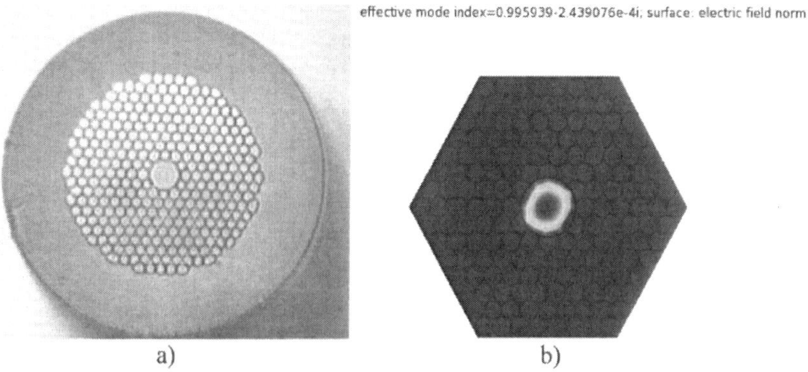

Figure 3. Hexagonal HC-PCF: a) scanning electron microscopy (SEM) image, with the hole diameter of 3.9μm and pitch of 4.0 μm; b) numerical result for the fundamental mode, when this fibre is filled with air, at 1atm.

Many applications, like nonlinear optics and data transmission, need a usable bandwidth as broad as possible. Since PBG HC-PCFs can only guide at frequencies where the PBG exists, it becomes important to study the possibility of increasing the width of the band gap. Several numerical studies have shown that this objective can be achieved by increasing the cladding air-filling fraction [22, 23]. Actually, a higher fraction of air in the periodic unit cell means thinner interconnecting struts, which are responsible for the distortion of all band gaps.

2.2. Inhibited Coupling HC-PCF

A different type of optical guidance mechanism has been found that does not require neither a higher index core material nor a cladding photonic bandgap. In this case, the fields of the core mode and of the cladding mode present little spatial overlap or a strong mismatch in their respective transverse spatial-phase. The guidance mechanism in this case is called inhibited coupling (IC) and was first experimentally observed with the introduction of Kagome lattice HC-PCF [24], Actually, the guided core-modes in Kagome HCPCF are longitudinally phase-matched with a cladding mode-continuum. However, coupling between the core-mode and the cladding modes is inhibited by two reasons. First, there is a strong transverse phase mismatch between them. Second, there is a weak spatial intersection between the core mode and the cladding modes due to a good confinement of the optical field in the core and a strong localization of the cladding modes in the cladding glass web.

Figure 4 shows (a) the crystal lattice and (b) the mode field profile of a 1-cell-core kagomé lattice HC-PCF with a pitch of 24 μm and a strut thickness of 0.2 μm. A good confinement of the optical field in the core is observed.

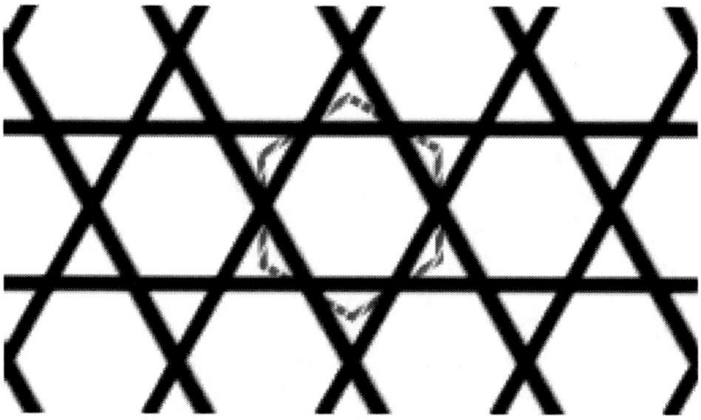

(a)

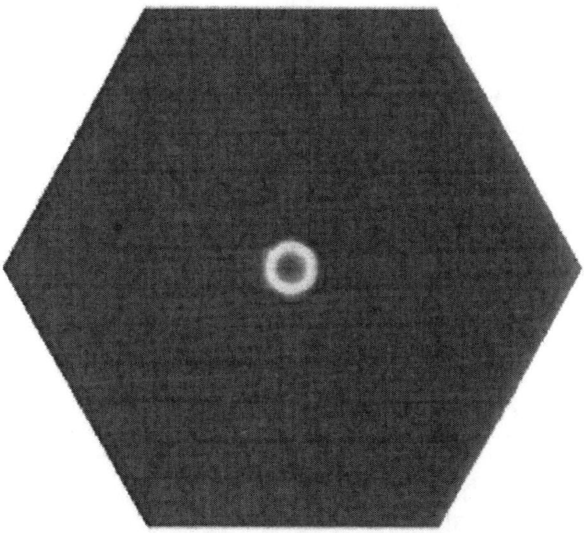

(b)

Figure 4. (a) Crystal lattice of a kagomé HC-PCF (the dashed hexagon shows the unit cell). (b) Mode field distributions at wavelength 600nm of a 1-cell-core kagomé lattice HC-PCFs with a pitch of 24 μm and a strut thickness of 0.2 μm.

2.3. HC-PCF Losses

Initial work on HC-PCFs was greatly stimulated by the possibility of HC-PCFs to provide ultimate ultra-low losses, well below the limits imposed by Rayleigh scattering in conventional all-glass fibers. However, the low air filling factor achieved in earlier HC-PCFs only allowed losses at the ~dB/cm level. Fibers with loss around ~1 dB/m were soon reported after the improvement of fibre properties [25]. Significantly better results were achieved by increasing the air filling fraction to levels above 90%, resulting in a two orders of magnitude reduction in transmission loss (13 dB/km at 1500 nm) [26]. A further increase in the air filling fraction provided a reduction of the loss in a 7c HC-PCF to 9.5 dB/km [27].

Several low-loss HC-PCFs were experimentally implemented by exploring different square and honeycomb lattices, as well as large pitch Kagome lattices [28, 29]. In all these cases the confinement loss remained above 300 dB/km because of the unavoidable strut connections or nodes. A much more significant reduction of the confinement loss in IC-HC-PCFs was achieved with the introduction of a hypocycloid core-shape, also known as negative curvature core-contour [30-32]. In this way, the confinement loss has been reduced from the 1 dB/m level to 8.5–70 dB/km [33-35]. It has been shown that the hypocycloid shape provides an enhancement of the inhibited coupling effect as a result of the decrease in the spatial overlap between the slow-varying transverse-phase core modes and the fast-varying transverse cladding modes [30, 36]. The relationship between the negative curvature and the confinement loss in the HC-PCFs has been experimentally and theoretically investigated by several authors [36,-46].

3. Dispersion and Nonlinearity of Gas-Filled Kagomé PCFs

In order to find the modes supported by a kagomé MOF, we can solve the following master equation [47]:

$$\nabla \times \left(\frac{1}{n^2(\lambda, r)} \nabla \times \mathbf{H}(\lambda, r) \right) = k_0^2 \mathbf{H}(\lambda, r); \tag{1a}$$

$$\mathbf{E}(\lambda, r) = \frac{i}{\omega \varepsilon_0 n^2(\lambda, r)} \nabla \times \mathbf{H}(\lambda, r); \tag{1b}$$

where $n(\lambda, r)$ is the refractive index at each point of the fiber's cross section, λ is the wavelength, **H** is the magnetic field and **E** is the electric field. Eq.(1a) and (1b) can be solved on two different domains: i) considering the whole cross section of the MOF or ii) only on a cell of its cladding and applying periodic boundary conditions. With these two approaches we obtain different, but consistent, information about the fibres: with the former method we obtain the propagation modes in the core, and optical properties such as dispersion and the nonlinear parameter; with the latter method we obtain the properties of the cladding structure such as the bandgap regions. Eqs. (1a) and (1b) have benn solved using COMSOL Multiphysics software [48].

The effective modal refractive index, n_{eff}, of kagomé HC-PCFs can be approximated by that of a glass capillary, which is given by [49]:

$$n_{eff} = n_{gas} - \frac{1}{2}\left(\frac{u_{01}\lambda}{\pi d_f}\right)^2, \tag{2}$$

where d_f is the core-diameter, u_{01} is the first zero of the J_0 Bessel function, λ is the wavelength, and n_{gas} is the refractive index of the core-filling gas, which is given by [50]:

$$n_{gas} = \sqrt{1 + \frac{p}{p_0}\left(\frac{b_1\lambda^2}{\lambda^2 - c_1} + \frac{b_2\lambda^2}{\lambda^2 - c_2}\right)}. \tag{3}$$

In Eq. (3), p is the gas pressure, $p_0 = 1$ bar, and the coefficients b_i, c_i depend on the considered gas (view table 1).

Table 1. Sellmeier coefficients for various gases (values taken from ref. [50])

	argon	helium	xenon
b1	b1 = 20332.29*1e-8	4977.77*1e-8	103701.61*1e-8
c1	c1 = 206.12*1e-6	28.54*1e-6	12.75*1e-6
b2	b2 = 34458.31*1e-8	1856.94*1e-8	31228.61*1e-8
c2	c2 = 8.066*1e-3	7.760*1e-3	0.561*1e-3

The propagation constant β of an optical pulse is related with the effective modal refractive index by:

$$\beta(\omega) = \frac{\omega n_{eff}}{c} \tag{4}$$

Mathematically, the effects of fiber dispersion are accounted for by expanding $\beta(\omega)$ in a Taylor series about the carrier frequency ω_0 at which the pulse spectrum is centred:

$$\beta(\omega) = \beta_0 + \frac{d\beta}{d\omega}(\omega - \omega_0) + \frac{1}{2}\frac{d^2\beta}{d\omega^2}(\omega - \omega_0)^2 + \ldots \tag{5}$$

where $\beta_0 = \beta(\omega_0)$,

$$\frac{d\beta}{d\omega} \equiv \beta_1 = \frac{1}{c}\left[n + \omega\frac{dn}{d\omega}\right] = \frac{n_g}{c} = \frac{1}{v_g} \tag{6}$$

and

$$\frac{d^2\beta}{d\omega^2} \equiv \beta_2 = -\frac{dv_g/d\omega}{v_g^2} \tag{7}$$

In Eq. (6), $n_g = n + \omega(dn/d\omega)$ and $v_g = c/n_g$ represent the group refractive index and the group velocity, respectively. The parameter β_2 characterizes the *group velocity dispersion* (GVD).

The kagomé HC-PCF provides ultra-broadband guidance at low loss levels and it presents, when evacuated, weak anomalous GVD over the entire transmission window. However, when filled with a noble gas, the normal GVD of the gas can be balanced against the anomalous GVD of the fiber, allowing the zero dispersion wavelength (ZDW) to be tuned across the ultraviolet, visible and near-infrared spectral regions. This can be observed from Figure 5, which shows the dependence of the GVD parameter β_2 on the wavelength for a kagomé HC-PCF with a 40□m core diameter filled with argon at various pressures.

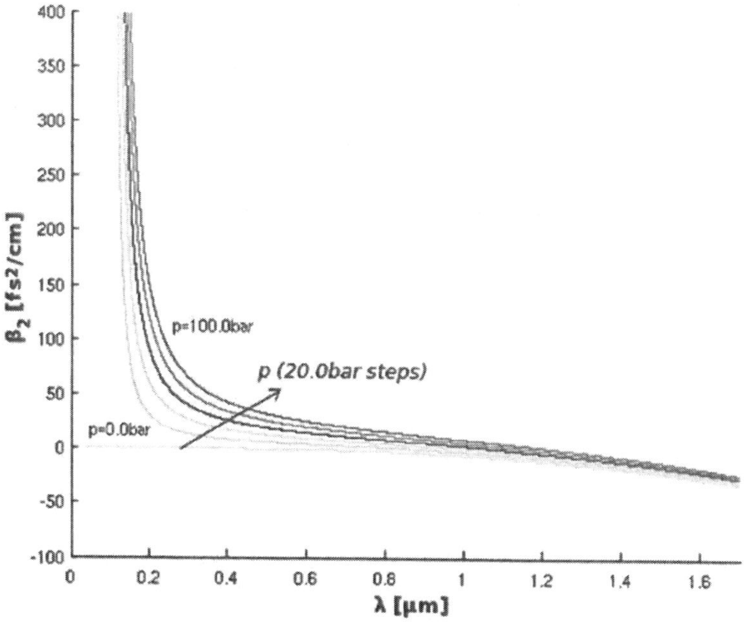

Figure 5. Dispersion curves of a kagome PCF with a 40μm core diameter, filled with 0 to 100 bar Ar.

We can compare the above results with those of a solid core PCF, where the shortest ZDW that can be achieved is about 500nm. Moreover, the dispersion magnitude is significantly smaller in a kagome PCF, which means that ultrafast pulses broaden much less quickly in this case.

Increasing the pressure in a gas-filled kagomé PCF results in an enhancement of the Kerr nonlinearity, that can approach that of silica glass. However, the dispersion remains low and flat from the UV to the near-IR. These features allow exploration of a range of different nonlinear regimes at relatively low pulse energies, without changing the fiber or the pump laser parameters.

Figure 6 shows the ZDW as a function of pressure for the fundamental mode of a kagomé HC-PCF with a core diameter of 20 μm filled with three different gases: xenon, argon, and helium. As the pressure is raised from 1 to 150 bar, the ZDW tunes smoothly from ~400 to 1400 nm in the case of xenon, whereas it varies from 300 to 950 nm for a argon filled core. Considering a pump wavelength at 790 nm, the dispersion in this case is anomalous (normal) below (above) 65.8 bar. However, in the case of a helium filled core and considering the same pump wavelength, the dispersion is anomalous for any pressure in the range 1 - 150bar.

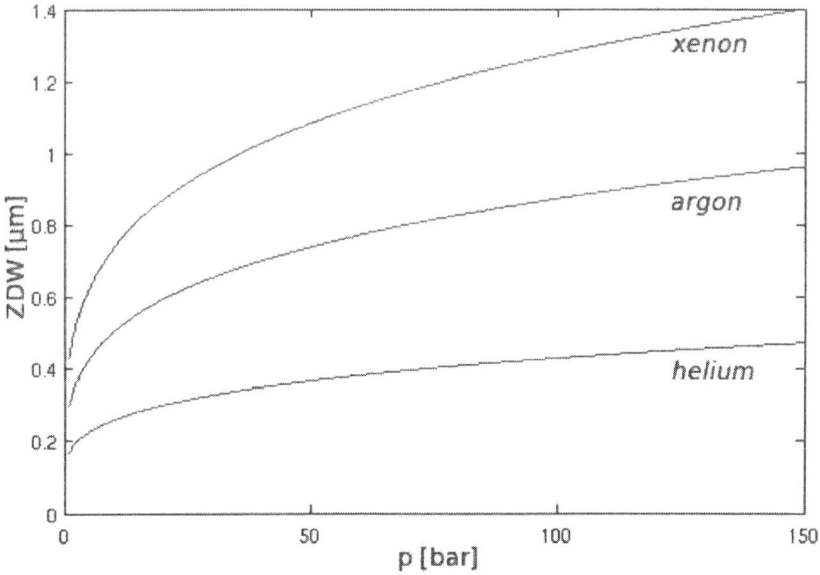

Figure 6. Zero dispersion wavelength (ZDW) as a function of pressure for the fundamental mode of a kagomé HC-PCF with a core diameter of 20 μm filled with three different gases: xenon, argon, and helium.

The nonlinear Kerr coefficient, $n_{Kerr}(p)$, of a gas-filled HC.PCF depends on the gas pressure as $n_{Kerr}(p) = p\, n^0_{Kerr}$, n^0_{Kerr} being the Kerr coefficient at 1atm [51]. Table 2 shows n_{Kerr} at 1atm for different noble gases.

Table 2. Kerr coefficients for various noble gases at pressure p = 1atm (after ref. [52])

$n_{kerr} x 10^{16}$ [m^2/W]

helium	neon	argon	krypton	xenon
$5.21x10^{-9}$	$1.31x10^{-8}$	$1.27x10^{-7}$	$3.07x10^{-7}$	$9.16x10^{-7}$

The nonlinear refractive index of argon becomes ~15 times smaller than that of bulk silica at 150 bar. In the case of xenon it becomes nearly half that of silica at 150 bar.

4. Ultrafast Nonlinear Dynamics

The propagation of a pulse through the kagomé HC-PCF can be described by the following generalized nonlinear Schrödinger equation (GNLSE) [53, 54]:

$$\frac{\partial u(z,t)}{\partial z} - i\sum_{k>1}\frac{1}{k!}i^k\beta_k\frac{\partial^k}{\partial\tau^k}u(z,t) = \gamma\left(1 + i\tau_{shock}\frac{\partial}{\partial\tau}\right)u(z,t)|u(z,t)| \tag{8}$$

where u is the normalized amplitude of the optical field, z is the distance along the fibre, $\tau_{shock} = 1/\omega_0$ is a timescale associated with effects such as the self-steepening and/or the optical shock formation, and the coefficient

$$\beta_k = \left.\frac{d^k\beta}{d\omega^k}\right|_{\omega=\omega_0} \tag{9}$$

represent the effects of the dispersion. In Eq. (8), γ is the fibre's nonlinear parameter, given by [54, 55]:

$$\gamma = \frac{2\pi n_{Kerr}(p)}{\lambda A_{eff}} \tag{10}$$

where A_{eff} is the effective mode area and $n_{Kerr}(p)$ is the nonlinear Kerr coefficient

Eq. (8) is written in the time domain. Alternatively, it can written in the frequency domain as:

$$\frac{\partial\tilde{u}(z,\omega)}{\partial z} - i[\beta - \beta_0 - \beta_1.(\omega - \omega_0)]\tilde{u}(z,\omega) = i\gamma\left(1 + \frac{\Delta\omega}{\omega_0}\right)F\{u(z,t)|u(z,t)|^2\} \tag{11}$$

where $\tilde{u}$ is the Fourier transform of u, and F represents the Fourier transform. Eq. (11) is theoretically identical to the common Eq. (8) for infinite number of β_k terms. However, solving Eq. (11) provides more accurate results than solving Eq. (8) for finite β_k terms.

4.1. Soliton-Effect Compression

When the input pulse propagates in the anomalous-GVD regime of the fiber, it becomes compressed through an interplay between SPM and GVD. This compression mechanism is related to the periodic evolution of the higher-order solitons, which go through an initial narrowing phase at the beginning of each period. If the fiber length is suitably chosen, the input pulses can be compressed by a factor that depends on the soliton order, N, given by [55]

$$N^2 = \frac{\gamma P_0 t_0^2}{|\beta_2|} \tag{12}$$

where P_0 and t_0 are the soliton peak power and width, respectively.

The optimum pulse compression factor, F_{opt}, of a soliton-effect compressor can be estimated from the following empirical relations [55]:

$$F_{opt} \approx 4.1N \tag{13}$$

In practice, extreme pulse compression is limited by higher order effects, namely by higher order dispersion. However, this limitation becomes less significant in the case of kagomé MOFs filled with noble gases, since they have a relatively smaller dispersion slope and they do not present Raman response. Compressed pulses with duration of some few fs can be achieved. Soliton-effect self-compression of ~30 fs pulses to shorter than 10 fs was achieved in an Ar-filled kagomé MOF in a 2011 experiment [56].

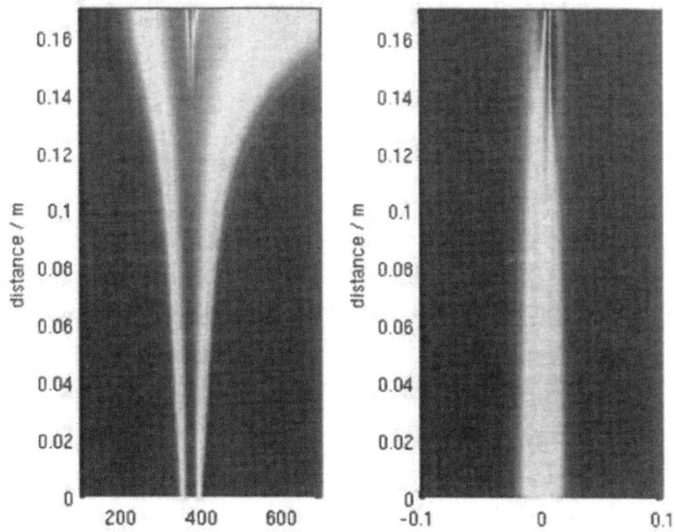

(a)

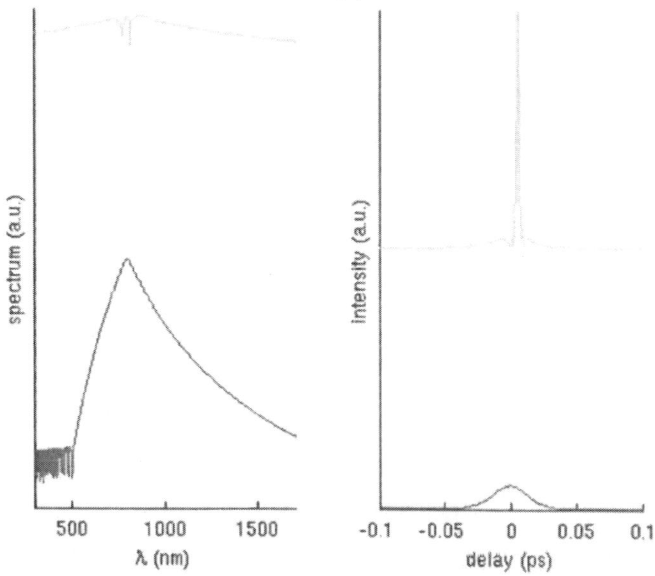

(b)

Figure 7. (a) Spectral and temporal evolution of self-compression of a 30 fs input pulse at 800nm through a kagomé PCF with a 30μm diameter core filled with Ar (ZDW at 500 nm). (b) Initial (lower part) and final (upper part) pulse profiles in spectral and time domains.

Figure 7 shows the spectral and temporal evolution of a 30fs pulse through a kagomé MOF with a 30μm diameter core filled with Ar, presenting a ZDW at 500 nm. The pumping is realized at 800nm, situated in the anomalous dispersion region, and the corresponding soliton order is N = 3.5. As a consequence of the soliton self-compression effect, the pulse temporal profile is dramatically sharpen, producing a ~2 fs pulse. For higher values of N the self-compressed pulses can achieve subcycle durations, but the corresponding quality factor is reduced [57].

The use of air-filled HC-MOFs has been also considered in recent years [58-60]. Actually, the ambient air has an almost two orders of magnitude lower nonlinearity than silica glass [61]. However, its nonlinearity can still become significant in the case of high peak power propagating pulses. The involved nonlinear processes can be exploited for pulse modification [62], new frequency generation [63], or even broad SG with a very high-power density due to the very high damage threshold of HC-MOFs and the confined gas within the core [64]. The simplicity of air-filled HCFs, combined with their very long light-matter interactions, motivated the exploration of potential applications in recent years [39].

The dispersion of a HC-MOF filled with ambient air is almost the same if air is replaced with 0.8 bar of Ar [60]. On the other hand, the Kerr coefficients at 400 nm in both cases are also very similar [65]. The main difference comes from the Raman effect, which is absent in the case of Ar and that is beneficial for spectral broadening and pulse compression in the case of ambient air [66].

In a 2021 experiment, 54-fs pulses at 400 nm were launched into a 80 cm-long air-filled kagomé PCF with a 22 μm core diameter [60]. Pump pulses with the energy of 2.6 μJ and duration of 54 fs were compressed temporally by a factor of 5, to a duration of ~11 fs. The experimental results demonstrated that both Raman and Kerr effects play a role in the compression dynamics.

4.2. UV-Light Generation

Extreme soliton-effect pulse compression of the input pulse results in a spectral expansion that overlaps with resonant dispersive-wave frequencies, which are consequently excited in the UV region. The resonance condition is given by:

$$\beta(\omega) - \beta_0 - \beta_1[\omega - \omega_0] = \frac{\gamma P_c}{2}, \tag{14}$$

where P_c is the self-compressed pulse peak power.

Figure 8 shows the spectral and temporal evolution of ultrafast nonlinear UV-generation in a kagomé HC-PCF with d = 40.0μm, and thickness = 0.1μm, filled with helium with p = 10bar. The laser signal has λ_{pump} = 700nm, FWHM of 40fs, and energy = 20mJ. The UV generation efficiency is defined as the fraction of the total output power at UV wavelengths (i.e., for wavelengths from 350nm to 10nm) compared to the total input power. In the case of Figure 8, the conversion efficiency is 6.9%. Another characteristic parameter is the quality of the UV emission, given by the ratio between the power within the full width at half maximum (FWHM) of the strongest UV peak and the total spectral power in the UV region. In the case of Figure 4, the quality factor is 84.1%.

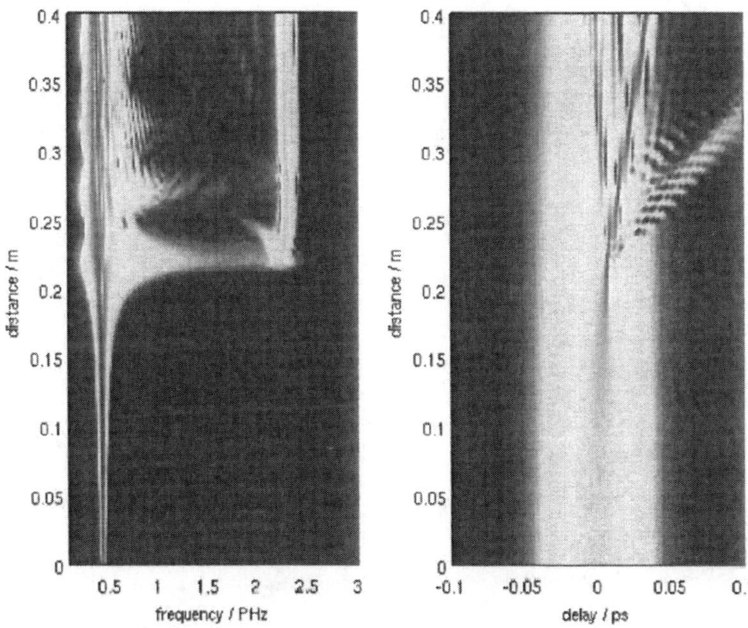

Figure 8. Spectral (left) and temporal (right) evolution of ultrafast nonlinear UV-generation in a kagomé HC-PCF with d = 40.0μm, and thickness = 0.1μm, filled with helium at pressure p = 10bar; the laser signal has λ_{pump} = 700nm, FWHM of 40fs, and energy = 20mJ.

In Figure 8 we can observe that the UV light is generated when the pulse achieves the maximum compression. After the maximum temporal compression and in consequence of the higher order dispersion effects, the pulse suffers solitonic fission, at a propagation distance, l_{fiss}, which is approximately given by:

$$l_{fiss} \sim l_d/N \, , \tag{15}$$

where $l_d = t_0^2/|\beta_2|$ is the dispersion length, and N is the soliton order.

Reducing the energy of the pump pulse to 15mJ, while maintaining the other parameters as in Figure 8, reduces the soliton order N. As a consequence, the fission length is increased, as observed in Figure 9. In this case, both the conversion efficiency and the quality factor are reduced to 4.3%, and 73.7%, respectively.

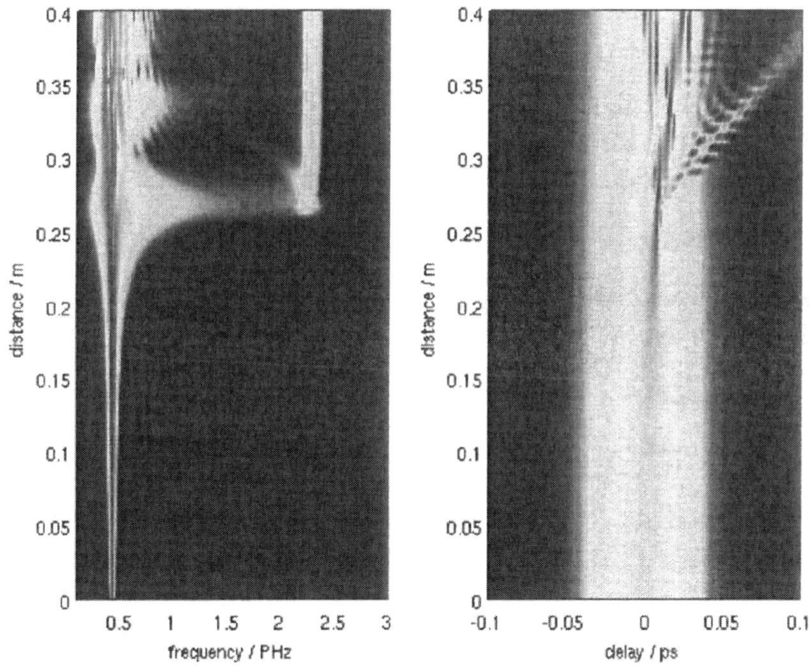

Figure 9. Spectral evolution in a kagomé HC-PCF with the same parameters as in figure 4, except that the pulse energy is15mJ.

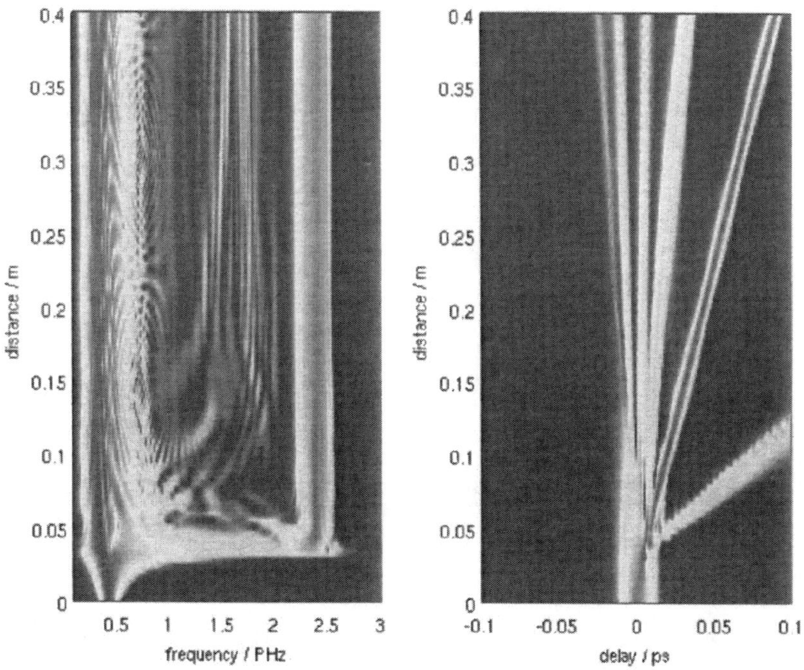

Figure 10. Spectral evolution in a kagomé HC-PCF with the same parameters as in figure 8, except that the input pulse FWHM is 10 fs.

Figure 10 shows the effects of changing the initial pulse width, while keeping the other parameters as in Figure 4. Reducing the pulse FWHM to 10 fs provides an enhancement of the conversion efficiency to 18.1%, whereas the quality factor of the UV radiation is slightly reduced to 75.1%. The fission distance decreases for smaller pulse widths, as expected.

Increasing the gas pressure to p = 15bar, while keeping the other parameters as in Figure 8, causes a shift of the central frequency of the UV light to lower values, as observed from Figure 11. On the other hand, the quality factor is reduced to 66.2%, whereas the efficiency of conversion is increased to 8.1%. An opposite effect can be observed reducing the gas pressure to p = 5 bar (Figure 12). In this case the central frequency of the UV light is shifted to higher values, whereas the quality factor is increased to 82.5%, and the efficiency of conversion is reduced to 5.4%. We conclude from Figures 8 to 12 that the frequency of the UV can be tuned by changing the pressure of the gas or the input pulse parameters.

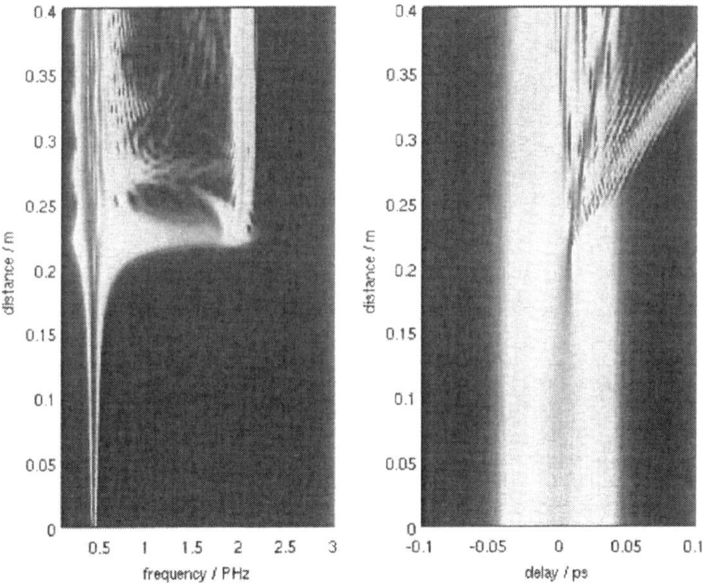

Figure 11. Spectral evolution in a kagomé HC-PCF with the same parameters as in figure 4, except for the gas pressure which is 15bar.

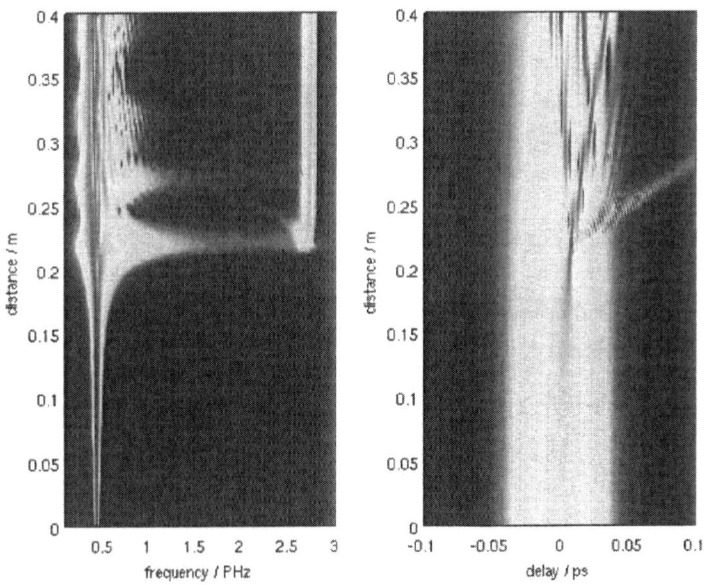

Figure 12. Spectral evolution in a kagomé HC-PCF with the same parameters as in figure 8, except for the gas pressure which is 5bar.

By comparing Figure 12 and Figure 8 we observe that by increasing the gas pressure, the central frequency of the UV light decreases In Figure 13, we confirm this behavior by considering several values for the gas pressure. When we increase the gas pressure the fiber's nonlinearity is increased, affecting also the dispersion profile. We conclude that the frequency of the final UV pulse can be tuned by changing the pressure of the gas in the fiber's core.

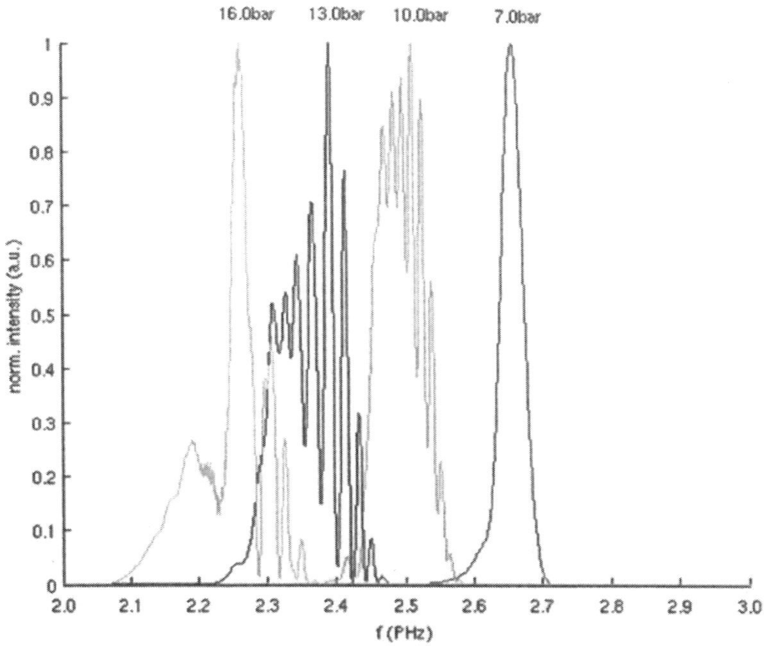

Figure 13. Output spectral profile of UV light generated in a kagomé HC-PCF; all the parameters are the same as in Figure 8 except the pressure of the gas, which is indicated in the label.

Conclusion

In this chapter we discussed the guiding mechanisms of both photonic bandgap and inhibited coupling hollow-core photonic crystal ibers (HC-PCFs). A special attention was paid to kagomé HC-PCFs, which cam provide lower losses and broadband guidance. These characteristics allow the possibility of studing the ultrafast dynamics of femtosecond pulse propagation in a gas-filled kagome HC-PCFs. By varying the gas pressure, the normal

GVD of the filling gas can be balanced against the anomalous GVD of the kagomé PCF allowing the zero dispersion wavelength to be tuned across the ultraviolet, visible and near-infrared spectral regions. Moreover, it was shown that the effective Kerr nonlinearity of a kagomé HC-PCF filled with xenon gas becomes of the same order of that of bulk silica when the pressure is increased from 1 to 150 bar. A generalized Schrödinger equation has been used to describe ultrashort pulse propagation in a gas-filled kagome PCF. The soliton self-compression effect has been observed, providing pulses with some few fs. Such extreme self-compression can lead to a highly efficient deep-UV dispersive wave generation. An ultrafast and coherent UV light source can be constructed using a kagomé HC-PCF, that is tunable by varying the gas pressure or the input pulse parameters.

Founding

Fundação para a Ciência e a Tecnologia (FCT) (Project UID/CTM/50025/2020).

References

[1] Cregan, R. F.; Mangan, B. J.; Knight, J. C.; Birks, T. A.; St. J. Russell, P. (1999). Single-Mode Photonic Band Gap Guidance of Light in Air. *Science 285*, 1537–1539.

[2] Hassan, M. R. A.; Yu, F.; Wadsworth, W. J.; Knight, J. C. (2016). Cavity-based mid-IR fiber gas laser pumped by a diode laser. *Optica. 3*, 218–221.

[3] Sollapur, R.; Kartashov, D.; Zurch, M.; Hoffmann, A.; Grigorova, T.; Sauer, G.; Hartung, A.; Schwuchow, A. Bierlich, J.; Kobelke, J.; Chemnitz, M., Schmidt, M. A., & Spielmann, C. (2017). Resonance-enhanced multi-octave supercontinuum generation in antiresonant hollow-core fibers. *Light Sci. Appl.*, *6*, e17124.

[4] Russell, P. S. J.; Holzer, P.; Chang, W.; Abdolvand, A.; Travers, J. C. (2014). Hollow-core photonic crystal fibres for gas-based nonlinear optics. *Nat. Photonics 8*, 278–286.

[5] Ouzounov, D. G.; Ahmad, F. R.; Muller, D.; Venkataraman, N.; Gallagher, M.T.; Thomas, M.G.; Silcox, J.; Koch, K.W.; Gaeta, A.L. (2003). Generation of megawatt optical solitons in hollow-core photonic band-gap fibers. *Science* 301, 1702–1704.

[6] Nissen, M.; Doherty, B.; Hamperl, J.; Kobelke, J.; Weber, K.; Henkel, T.; Schmidt, M. A. (2008). UV Absorption Spectroscopy in Water-Filled Antiresonant Hollow Core Fibers for Pharmaceutical Detection. *Sensors 18*, 478.

[7] Heckl, O. H.; Saraceno, C. J.; Baer, C. R.; Sudmezer, T.; Wang, Y. Y.; Cheng, Y.; Benabid, F.; Keller, U. (2011). Temporal pulse compression in a xenon-filled kagome-type hollow-core photonic crystal fiber at high average power. *Opt. Express*, *19*, 19142–19149.

[8] Russell, P. St. J. (2006). Photonic-crystal fibers. *J. Lightwave Technol.* 24(12): 4729-4749.

[9] Couny, F., Benabid, F., Roberts, P., Light, P., and Raymer, M. (2007). Generation and photonic guidance of multi-octave optical-frequency combs *Science* 318 (5853), 1118–1121

[10] Pearce, G., Wiederhecker, G., Poulton, C., S. Burger, and Russell, P. (2007). *Opt. Express* 15(20), 12680–12685.

[11] Joly, N., Nold, J., Chang, W., Hölzer, P., Nazarkin, A., Wong, G., Biancalana, F., and Russell, P. (2011). "Bright spatially coherent wavelength-tunable deep-UV laser source using an Ar-filled photonic crystal fiber," *Phys. Rev. Lett.*, 106(20): 203901:203904.

[12] Mak, K., Travers, J., Hölzer, P., Joly, N., and Russell, P. (2013). Tunable vacuum-UV to visible ultrafast pulse source based on gas-filled kagome-PCF, *Opt. Express*, 21(9):10942-10953.

[13] Russell P. St. J. (2003). Photonic crystal fibers. *Science*, 299, 358-362.

[14] Couny F., Benabid F., Roberts P. J., Light P. S., and Raymer M. G. (2007). Generation and photonic guidance of multi-octave optical-frequency combs. *Science*, 318, 1118-1121.

[15] Benabid F, Roberts P. J. Linear and nonlinear optical properties of hollow core photonic crystal fiber. *J. Mod. Optics* 2011;58:87–124.

[16] Birks T. A., Pearce G. J., Bird D. M. Approximate band structure calculation for photonic bandgap fibres. *Opt. Express* 2006;14:9483–90.

[17] West J. A, Fajardo J., Gallagher M., Koch K., Borrelli N., Allan D. Demonstration of an IR-optimized air-core photonic band-gapfiber. *European Conference on Optical Communication (ECOC)* 2000, paper Th.A.2.

[18] Saitoh K., Koshiba M. Leakage loss and group velocity dispersion in air-core photonic bandgap fibers. *Opt. Express* 2003;11:3100–9.

[19] White T. P., McPhedran R. C., Botten L. C., Smith G. H., de Sterke C. M. Calculations of air-guided modes in photonic crystal fibers using the multipole method. *Opt. Express* 2001;9:721–32.

[20] Aghaie K. Z, Fan S. H, Digonnet M. J. F. Birefringence Analysis of photonic-bandgap fibers using the hexagonal yee's cell. *IEEE J. Quantum. Elect.* 2010;46: 920–30.

[21] Couny, F.; Benabid, F.; Roberts, P. J.; Burnett, M. T.; Maier, S. A. (2007). Identification of Bloch-modes in hollow-core photonic crystal fiber cladding. *Opt. Express*, *15*, 325–338.

[22] Mortensen N. A, Nielsen M. D. (2004). Modeling of realistic cladding structures for air-core photonic bandgap fibers. *Opt. Lett.*; 29:349–51.

[23] Saitoh K, Koshiba M. (2003). Leakage loss and group velocity dispersion in air-core photonic bandgap fibers. *Opt. Express*; 11:3100–9.

[24] Benabid, F.; Knight, J. C.; (2002). Antonopoulos, G.; Russell, P. S. J. Stimulated Raman scattering in hydrogen-filled hollow-core photonic crystal fiber. *Science*, *298*, 399–402.

[25] West J. A., Venkataramam N., Smith C. M., Gallagher M. T. Photonic crystal fibers. *European Conference on Optical Communications (ECOC)* 2001, paper Th.A.2.2.

[26] Smith C. M., Venkataraman N., Gallagher M. T., Muller D., West J. A., Borrelli N. F, Allan D. C., Koch K. W. (2003). Low-loss hollow-core silica/air photonic bandgap fibre. *Nature*; 424:657–9.

[27] Amezcua-Correa R., Gerome F., Leon-Saval S. G., Broderick N. G. R., Birks T. A., Knight J. C. (2008). Control of surface modes in low loss hollow-core photonic bandgap fibers. *Opt. Express*; 16:1142–9.

[28] Couny, F.; Benabid, F.; Light, P. S.(2006). Large-pitch kagome-structured hollow-core photonic crystal fiber. *Opt. Lett. 31*, 3574–3576.

[29] Couny, F.; Roberts, P. J.; Birks, T. A.; Benabid, F. (2008). Square-lattice large-pitch hollow-core photonic crystal fiber. *Opt. Express 16*, 20626–20636.

[30] Wang, Y. Y.; Wheeler, N. V.; Couny, F.; Roberts, P. J.; Benabid, F. (2011). Low loss broadband transmission in hypocycloid-core Kagome hollow-core photonic crystal fiber. *Opt. Lett.*, *36*, 669–671.

[31] Pryamikov, A. D.; Biriukov, A. S.; Kosolapov, A. F.; Plotnichenko, V. G.; Semjonov, S. L.; Dianov, E. M. (2011). Demonstration of a waveguide regime for a silica hollow—Core microstructured optical fiber with a negative curvature of the core boundary in the spectral region > 3.5 μm. *Opt. Express 19*, 1441.

[32] Yu, F.; Wadsworth,W. J.; Knight, J. C. (2012). Low loss silica hollow core fibers for 3–4 μm spectral region. *Opt. Express 20*, 11153–11158.

[33] Bradley T. D., Wang Y., Alharbi M., Debord B., Fourcade-Dutin C., Beaudou B., Gerome F., and Benabid F. (2013). Optical properties of low loss (70dB/km) hypocycloid-core kagome hollow core photonic crystal fiber for Rb and Cs based optical applications. IEEE *J. Lightwave Technol.*, 31 (16), 2752-2755.

[34] Wheeler N. V., Bradley T. D., Hayes J. R., Gouveia M. A., Liang S., Chen Y., Sandoghchi S. R., Mousavi S. M. A., Poletti F., Petrovich M. N., and Richardson D. J. (2017). Low-loss kagome hollow-core fibers operating from the near- to the mid-IR. *Opt. Lett.*, 42 (13), 2571-2574.

[35] Maurel M., Chafer M., Amsanpally A., Adnan M., Amrani F., Debord B., Vincetti L., Gérôme F., and Benabid F. (2018). Optimized inhibited-coupling kagome fibers at Yb-Nd:Yag (8.5 dB/km) and Ti:Sa (30 dB/km) ranges. *Opt. Lett.*, 43 (7), 1598-1601.

[36] Debord B., Alharbi M., Bradley T. D., Fourcade-Dutin C., Wang Y. Y., Vincetti L., Gérôme F., and Benabid F. (2013). Hypocycloid-shaped hollow-core photonic crystal fiber Part I: arc curvature effect on confinement loss. *Opt. Express*, 21 (23), 28597-28608.

[37] Belardi W. and Knight J. C. (2014). Hollow antiresonant fibers with reduced attenuation. *Opt. Lett.*, 30 (7), 1853-1856.

[38] Belardi W. and Knight J. C. (2014). Hollow antiresonant fibers with low bending loss. *Opt. Express*, 22 (8), 10091-10096.

[39] Debord B., Alharbi M., Vincetti L., Husakou A., Fourcade-Dutin C., Hoenninger C., Mottay E., Gérôme F., and Benabid F. (2014). Multi-meter fiber-delivery and pulse self-compression of milli-Joule femtosecond laser and fiber-aided laser-micromachining. *Opt. Express*, 22 (9), 10735-10746.

[40] Poletti F. (2014). Nested antiresonant nodeless hollow core fiber. *Opt. Express*, 22 (20), 23807-23828.

[41] Hartung A., Kobelke J., Schwuchow A., Wondraczek K., Bierlich J., Popp J., Frosch T., and Schmidt M. A. (2014). Double antiresonant hollow core fiber-guidance in the deep ultraviolet by modified tunneling leaky modes. *Opt. Express*, 22 (16), 19131-19140.

[42] Jiang X., Joly N. Y., Finger M. A., Babic F., Wong G. K. L., Travers J. C., and Russell P. St. J. (2015). Deep-ultraviolet to mid-infrared supercontinuum generated in solid-core ZBLAN photonic crystal fibre. *Nature Photon*., 9, 133-139.

[43] Yu F. and Knight J. C. (2016). Negative curvature hollow-core optical fiber. IEEE J. *Selected Top. Quantum Electron*., 22 (2), 4400610.

[44] Wang Y. and Ding W. (2017). Confinement loss in hollow-core negative curvature fiber: A multi-layered model. *Opt. Express*, 25 (26), 33122-33133.

[45] Bufetov I. A., Kosolapov A. F., Pryamikov A. D., Gladyshev A. V., Kolyadin A. N., Krylov A. A., Yatsenko Y. P., and Biriukov A. S. (2018). Revolver hollow core optical fibers. *Fiber*s, 6, 39-1-39-26.

[46] Yu F., Song P., Wu D., Birks T., Bird D., and Knight J. C. (2019). Attenuation limit of silica-based hollow-core fiber at mid-IR wavelengths. *Appl. Phys. Lett. Photon*., 4, 080803-1-080803-6.

[47] Joannopoulos, J. D., Johnson, S. G., Winn, J. N., and Meade, R. D. (2008). *Photonic crystals: molding the flow of light*. Princeton University Press, 2nd edition.

[48] COMSOL Multiphysics – User's guide, v4.2.

[49] Im, S.-J., Husakou, A., and Herrmann, J. (2009). Guiding properties and dispersion control of kagome lattice hollow-core photonic crystal fibers. *Opt. Express*, 17(15):13050-13058.

[50] Börzsönyi, A., Heiner, Z., Kalashnikov, M. P., Kovács, A. P., and Osvay, K. (2008). Dispersion measurement of inert gases and gas mixtures at 800 nm *Applied Optics*, 47(27):4856-4863.

[51] Börzsönyi, A., Heiner, Z., Kovács, A. P., Kalashnikov, M. P., and Osvay1, K. (2010). Measurement of pressure dependent nonlinear refractive index of inert gases. *Optics Express*, 8, 25847-25854.

[52] Brée, C. (2012). *Nonlinear optics in the filamentation regime*. Springer Theses, Humboldt University of Berlin, Germany.

[53] Dudley J. M., and Taylor, J. R. (2010). *Supercontinuum generation in optical fibers*, Cambridge University Press; 1st edition.

[54] Ferreira, M. F. S. (2011). *Nonlinear effects in optical fibers*. John Wiley & Sons, Hoboken, New Jersey.

[55] Ferreira, M. F. S. (2022). *Solitons in optical fiber systems*. John Wiley & Sons, Hoboken, New Jersey.

[56] Joly, N., Nold, J., Chang, W., Hölzer, P., Nazarkin, A., Wong, G., Biancalana, F., and Russell, P. (2011). Bright spatially coherent wavelength-tunable deep-UV laser source using an ar-filled photonic crystal fiber. *Phys. Rev. Lett.* 106, 203901.

[57] Travers, J., Chang, W., Nold, J., Joly, N., and Russell. P. (2011). Ultrafast nonlinear optics in gas-filled hollow-core photonic crystal fibers. *J. Optical Soc. Am. B* 28, A11.

[58] Mousavi, S., Mulvad, H., Wheeler, N., Horak, P., Hayes, J., Chen, Y., Bradley, T., Alam, S.-U., Sandoghchi, S., Fokoua, E., Richardson, D., and Poletti, F. (2018). Nonlinear dynamic of picosecond pulse propagation in atmospheric air-filled hollow core fibers. *Opt. Express* 26, 8866 -8882.

[59] Debord, B., Maurel, M., Gérôme, F., Vincetti, L., Husakou, A., and Benabid, F. (2019). Strong nonlinear optical effects in micro-confined atmospheric air. *Photonics Res.* **7**, 1134-1141.

[60] Luan, J., Russell, P., and Novoa, N. (2021). Efficient compression of ultrashort near-UC pulses in air-filled hollow-core photonic crystal fibers (2021). *Optics Express* 29, 13787-13793.

[61] Mlejnek, M., Wright, E. M., and Moloney, J. V. (1998). Dynamic spatial replenishment of femtosecond pulses propagating in air. *Opt. Lett.* 23, 382-384.

[62] Guichard, F., Giree, A., Zaouter, Y., Hanna, M., Machinet, G., Debord, B., Gérôme, F., Dupriez, P., Druon, F., Hönninger, C., Mottay, E., Benabid, F., and Georges, P. (2015). Nonlinear compression of high energy fiber amplifier pulses in air-filled hypocycoid-core kagome fibers. *Opt. Express* 23, 7416-7423.

[63] Bhagwat, S., and Gaeta, A. (2008). Nonlinear optics in hollow-core photonic bandgap fibers *Opt. Express* 16, 5035-5047.

[64] Yatsenko, Y., Pleteneva, E., Okhrimchuk, A., Gladyshev, A., Kosolapov, A. Kolyadin, A., and Bufetov, I. (2017). Multiband supercontinuum generation in an air-core revolver fibre *Quantum Electron.* 47, 553-560.

[65] Zahedpour, S., Wahlstrand, J., and Milchberg, H. (2015) Measurement of the nonlinear refractive index of air constituents at mid-infrared wavelengths *Opt. Lett.* 40, 5794-5797.

[66] Belli, F., Abdolvand, A., Chang, W., Travers, J., and Russell, P. (2015), Vacuum-ultraviolet to infrared supercontinuum in hydrogen-filled photonic crystal fiber, *Optica* 2, 292-300.

Chapter 3

Nonlinear Processes in Photonic Crystal Fibers in Emerging Applications

Debashri Ghosh[1], PhD
Samudra Roy[2], PhD
and Shyamal Bhadra[1,*], PhD
[1]Fiber Optics and Photonics Division, CSIR-Central Glass and Ceramic Research Institute, Kolkata, India
[2]Department of Physics, Indian Institute of Technology - Kharagpur, Kharagpur, India

Abstract

Light propagates in all directions once it is created but can be directed with special technique. Light guidance in various types of optical fibers is well known in the linear domain when material property does not change with the modal intensity. However, situation gets complicated once intensity dominates over mode propagation, which gives an opportunity to study many facets of nonlinearity in guided media enabling an enormous range of technological advantages. The advent of matured techniques in the late 1980's to tailor structures on the micro and nanoscale provided the opportunity to investigate the interaction of light with structured matter. This framework led to the emergence of the photonic crystals when in 1987 Eli Yablonovitch [1] and Sajeev John [2] independently proposed the idea of photonic bandgap (PBG).

---

* Corresponding Author's Email: skbhadra88@gmail.com.

In: Photonic Crystal Fibers (PCFs)
Editor: Mário F. S. Ferreira
ISBN: 979-8-88697-618-2

With the idea of photonic bandgap in place and with the motivation for creating a new kind of dielectric waveguide that will guide light in a confined manner over distances that are much larger than the Rayleigh range, Philip Russell proposed the two-dimensional photonic crystal structure in 1996 in the form of photonic crystal fiber (PCF) [3]. The initial idea was to trap light in a hollow core of a glass fiber with the help of a 2-D photonic crystal formed by microscopic air holes running along the entire length of the fiber. The first successful silica-air photonic crystal fiber structure that was fabricated in 1996 had the hole-diameter-to-pitch ratio of ~0.2 that was theoretically too small for PBG guidance in a hollow core and so the final fiber was fabricated with a solid silica core surrounded by 216 air channels that ultimately led to the discovery of the endlessly single mode photonic crystal fiber [3, 4]. The initial success rapidly led to the development of a series of new types of PCFs and theoretical work also progressed in tandem.

In this chapter, following the chronology, we will briefly describe the guidance mechanism of conventional optical fiber, its properties and then move on to the fundamentals of the photonic crystal which will underline the development of the photonic crystal fibers. The two types of photonic crystal fibers – solid and hollow core, will then be dealt in details, starting from their light guiding mechanisms to their unique and interesting properties and finally to the applications that will also chronicle the recent developments in this field.

Keywords: Dirac mode, dispersion, hollow core PCF, kagomé fiber, Kerr nonlinearity, photonic bandgap, photonic crystal fiber (PCF), plasmonics

Introduction

The concept of telecommunication through optical mode was tried experimentally long back using various methods. Photophone by Graham Bell was perhaps the first demonstration of this kind of communication. Almost after hundred years a revolutionary prediction was made in November, 1965 in a paper titled *Dielectric-fiber surface waveguides for optical frequencies* by Kao and Hockham of Standard Telecommunication Laboratories Ltd., UK. The paper was actually published with a brief revision in the Proceedings of IEEE [5]. Transmission of huge data and information needs wide bandwidth of frequencies which is possible through optical signals, and now optical fibers constitute the global backbone of such connectivity. Payne and Gambling in

1975 at the Southampton University, UK had shown an interesting property of silica glass that the optical pulse spreading is zero at 1270 nm wavelength, which means material dispersion in silica is zero at that wavelength [6]. This zero dispersion wavelength (ZDW) can be manipulated depending on the doping materials used in various methods of fiber fabrication. In late '80s, the invention of optical amplifiers had boosted further the future projected demand for bandwidth of optical fiber telecommunication network.

In a single optical fiber multiple wavelengths can be transmitted through WDM process which needs more power and closely spaced optical channels. Therefore, in a single mode (SM) fiber small magnitude of material or chromatic dispersion is to be introduced in long haul WDM network to mitigate the effects of nonlinearity. The third order nonlinear effects such as cross phase modulation (XPM) and four wave mixing (FWM) degrade the propagating signals. At high power the fiber exhibits fast nonlinear response since the nonlinear dielectric polarization is proportional to the third power of the electric field of the propagating signal. A solution to this problem is to deploy non-zero dispersion shifted fiber (NZDSF) which is now commercially installed in long haul WDM network. Nonlinearity in telecom fiber creates problem in signal transmission as indicated above; on the other hand, this property can be utilized in specialty optical fiber for niche applications. In artificial photonic crystals where micrometric air holes are incorporated in a transparent dielectric, light can be trapped or guided, which suggests a radically new mechanism of light propagation as demonstrated independently by Eli Yablonovitch and Sajeev John in 1987. Extending this idea, photonic crystal fiber (PCF) was first demonstrated by Prof. Philip Russell and his group in 1996 [3], which is subdivided into two basic types - solid core PCF (SCPCF) and hollow core PCF (HCPCF) [7]. In a PCF, one can confine intense light in the core to exploit the various nonlinear processes that lead to the generation of broadband light, termed as supercontinuum, at the output. We will attempt to describe these nonlinear processes in detail in this chapter. Additionally, some applications of HCPCF will be mentioned to show the future trend of research with interesting applications.

Conventional Optical Fiber

In this section we briefly review the light guidance principle and important properties of conventional optical fibers.

Light Guidance Mechanism

Conventional optical fiber is basically a cylindrical dielectric waveguide that transmits light along its axis through the process of total internal reflection (TIR). The simplest optical fiber - the step index fiber (SIF) – consists of a dielectric core surrounded by another dielectric cladding. In SIF, the refractive index of the core n_c is made slightly higher than the refractive index of the cladding n_{cl} to ensure the TIR. The graded index (remove the hyphen) fiber is another kind of optical fiber where the refractive index of the core decreases gradually from centre to core boundary. In ray approach, exploiting the *Snell-Descartes law*, it is easy to see that in SIF, due to the condition $n_c > n_{cl}$, the propagating light ray in the core is totally reflected back from the core-cladding interface when the angle between the direction of propagation and the core-cladding interface is less than the critical angle. In fact, all the rays that fall on the fiber core making an angle less than the acceptance angle θ_a (defined by $\theta_a = \sin^{-1}\sqrt{n_c^2 - n_{cl}^2}$) with respect to the fiber axis should propagate through the fiber core. The process is illustrated in Figure 1(a). For infinitely long SIF, propagation constant β of the core guided mode satisfies the condition, $n_{cl}k_0 < \beta < n_c k_0$, where $k_0 = 2\pi/\lambda$ is the free space wave vector. The transverse wave number, $k_\perp = \sqrt{n_c^2 k_0^2 - \beta^2}$ for the guided modes ($\beta < n_c k_0$) is real and much smaller compared to the propagation constant β (see Figure 1(b)).

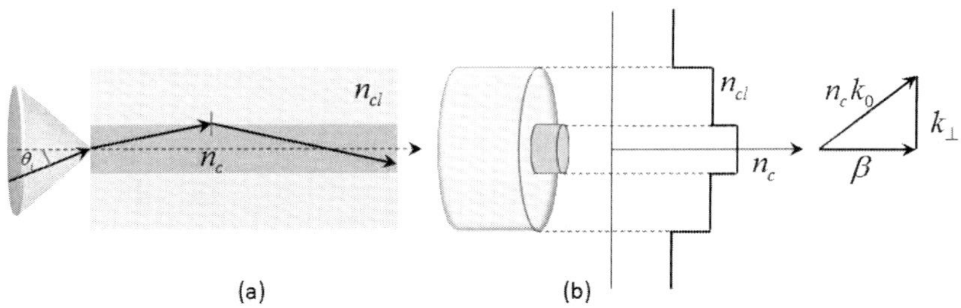

Figure 1. (a) Light acceptance cone and ray propagation mechanism in an optical fiber. (b) Refractive index variation in step index fiber.

Fiber Modes

For a given frequency ω, an optical fiber supports a finite number of guided modes whose spatial distribution $\tilde{\mathbf{E}}(\mathbf{r},\omega)$ is a solution of the wave equation,

$\nabla^2\tilde{\mathbf{E}} + n^2(\omega)k_0^2\tilde{\mathbf{E}} = 0$, where $k_0 = \omega/c$ and $\tilde{\mathbf{E}}(\mathbf{r},\omega)$ is the *Fourier transform* of the field $\mathbf{E}(\mathbf{r},t)$ defined as, $\tilde{\mathbf{E}}(\mathbf{r},\omega) = \int_{-\infty}^{\infty} E(r,t)e^{i\omega t}dt$. The field moving along the z-direction is generally expressed as, $\tilde{\mathbf{E}}(\mathbf{r},\omega) = A(\omega)F(\rho)e^{il\varphi}e^{i\beta z}$, where l takes the integer values $l = 0, 1, 2, \ldots$. In the core region ($\rho \leq a$), the radial distribution of the field $F(\rho)$ satisfies the well-known differential equation,

$$\frac{d^2F}{d\rho^2} + \frac{1}{\rho}\frac{dF}{d\rho} + \left(n^2k_0^2 - \beta^2 - \frac{l^2}{\rho^2}\right)F = 0 \tag{1}$$

whose general solutions are the *Bessel functions* $F(\rho) = C_1J_l(p\rho) + C_2N_l(p\rho)$. Here $J_l(x)$ and $N_l(x)$ are the Bessel and Neumann function respectively with $p = \sqrt{n_c^2k_0^2 - \beta^2}$. The singularity of $N_l(x)$ at $x = 0$ makes $C_1 = 0$ and we eventually get the physically meaningful solution

$$F(\rho) = C_1J_l(p\rho) \tag{2}$$

In the cladding region ($\rho \geq a$) we can have the exponentially decaying solution in the form of modified Bessel function as $F(\rho) = C_3K_l(q\rho)$, where C_3 is an arbitrary constant and $q = \sqrt{\beta^2 - n_{cl}^2k_0^2}$. The boundary condition that the core and cladding fields are continuous and smoothly varying at the core-cladding interface ($\rho = a$) leads to the following transcendental equation whose solutions determine the propagation constant β for the fiber modes [8],

$$\left[\frac{J_l'(U)}{UJ_l(U)} + \frac{K_l'(W)}{WK_l(W)}\right]\left[\frac{J_l'(U)}{UJ_l(U)} + \frac{n_c^2}{n_{cl}^2}\frac{K_l'(W)}{WK_l(W)}\right] = \\ = l^2\left(\frac{1}{U^2} + \frac{1}{W^2}\right)\left(\frac{1}{U^2} + \frac{n_c^2}{n_{cl}^2}\frac{1}{W^2}\right) \tag{3}$$

where $U = pa$ and $W = qa$ such that $V = \sqrt{U^2 + W^2} = k_0 a\left(n_c^{\,2} - n_{cl}^{\,2}\right)^{1/2}$, and prime denotes the differentiation with respect to the argument. The *eigen value equation*, Eqn. (3) has several solutions for $l = 0, 1, 2, \ldots$. and represented as β_{ml}, where $m = 1, 2, 3, \ldots$ defines the mth solution of β in decreasing order for a given l. For $l = 0$ the z-component of either the electric or magnetic field vanishes and the modes are analogous to the transverse electric (TE) and transverse magnetic (TM) modes of a planar waveguide. However for $l > 0$, fiber modes become hybrid and all six components of the electromagnetic field become non-vanishing. In Figure 2, spatial distributions of few modes are shown that are excited in a SIF. The number of modes supported by the SIF typically depends on the V-parameter. For example, only fundamental mode (represented as LP_{01}) is supported if $V \leq 2.405$. The fundamental mode in a fiber is typically defined as, $F(\rho) = J_0(p\rho)$ and can be approximated as Gaussian distribution $F(x,y) \approx e^{-(x^2+y^2)/r^2}$, where r is the spot size of the mode [9, 10].

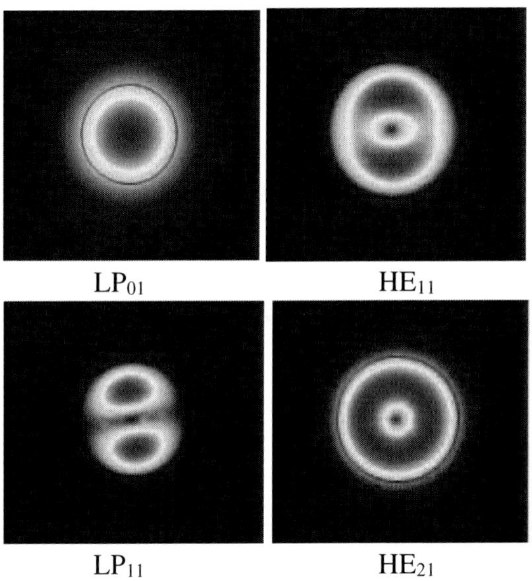

Figure 2. Different modes generated in a step-index fiber. LP_{01} in the top left denotes the fundamental mode while the other three (HE_{11}, LP_{11}, HE_{21}) are higher order modes in a SIF.

Fiber Attenuation

Fiber attenuation is an important fiber parameter which measures the amount of power loss during transmission of an optical signal inside the fiber. The fiber loss can be modeled by the *Beer's law* where the transmitted power P_t can be given as, $P_t = P_0 e^{-\alpha L}$, where P_0 is the input power, α is the fiber *attenuation coefficient* and L is the propagating length. It is customary to express α in unit of dB/km using the relation,

$$\alpha_{dB} = -\frac{10}{L}\log\frac{P_t}{P_0} = 4.343\alpha \qquad (4)$$

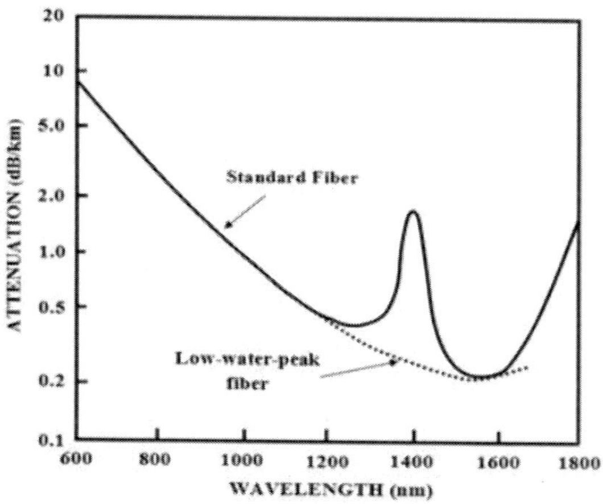

Figure 3. Spectral attenuation of a standard single mode optical fiber (bold line) having the OH absorption peak at around 1390 nm wavelength and the lowest loss of around 0.2 dB/km at 1550 nm. The dotted line denotes the attenuation loss of an optical fiber specially designed and fabricated to have extremely low water loss and hence no absorption peak at 1390 nm. (Redrawn from the reference [11].)

The complex refractive index $n_t(\omega) = n(\omega) + i\kappa(\omega)$ leads to the optical loss through the *extinction coefficient* κ. The *attenuation coefficient* α is related to κ as, $\alpha(\omega) = 2\omega\kappa/c$ showing a frequency (ω) dependence. Figure 3 shows the typical loss spectrum of a silica fiber fabricated by the MCVD process. The fiber exhibits minimum loss of about ~0.2 dB/km near 1.55 μm. Loss increases considerably at lower and higher wavelengths reaching a value of few dB/km in the visible region. However, even a 10 dB/km loss corresponds to $\alpha \approx 2\text{x}10^{-5}$ cm^{-1}, which is much lower compared to that of most other materials. Material absorption and *Rayleigh scattering* are the dominant contributors to the loss spectrum. Silica based glass is mostly transparent in the wavelength range 0.5 μm to 2 μm whereas it has electronic resonances in the ultraviolet region, and vibrational resonances in the far-infrared region beyond 2 μm. In practice OH ion is the most important impurity that affects fiber loss owing to its fundamental vibrational absorption peak at ≈2.73 μm, overtone of which is responsible for the dominant peak near 1.4 μm and a smaller peak near 1.23 μm respectively. The density fluctuation that freezes into the fused silica during manufacture yields *Rayleigh scattering*, the fundamental loss mechanism in fiber especially at low wavelengths. Under the Rayleigh scattering loss which varies as λ^{-4}, the light is scattered in all directions. The Rayleigh scattering loss is intrinsic to the fiber and estimated by the coefficient $\alpha_R = K_R \lambda^{-4}$, where the constant K_R typically ranges between 0.7-0.9 dB/km-μm^4 depending on the constituents of the fiber core. Near the operating wavelength of 1.55 μm, α_R is typically 0.12-0.15 dB/km, which can be reduced to about 0.05 dB/km in some glasses.

Fiber Dispersion

The electromagnetic wave propagating through a dielectric medium interacts with the bound electrons and as a result the medium responds which is, in general, frequency (ω) dependent. The response of the medium manifests through the frequency dependence of the refractive index $n(\omega)$, the origin of *chromatic dispersion*. In general, dispersion originates partly from the frequency dependent response of the material to the electromagnetic waves (by the *Sellmeier equation*), and partly from mode confinement associated with the geometry of the waveguide resulting in the effective refractive index $n_{eff}(\omega)$. These two components are referred to as material and waveguide dispersion respectively and combine to form the total dispersion. The variation of the refractive index with frequency is modeled by the *Sellmeier equation*,

$$n^2(\omega) = 1 + \sum_{j=1}^{m} \frac{B_j \omega_j^2}{\omega_j^2 - \omega^2} \tag{5}$$

where, ω_j is the resonance frequency and the constant B_j represents the strength of the jth resonance. The dispersion plays a crucial role in pulse propagation inside an optical fiber where different spectral components associated with the pulse travel at different speeds given by $c/n_{eff}(\omega)$.

The effect of dispersion is taken into consideration by expanding the mode propagation constant $\beta(\omega)$ [$= n_{eff}(\omega)\omega/c$] in a Taylor series around an operating frequency ω_0,

$$\beta(\omega) = \beta(\omega_0) + \beta_1(\omega_0)(\omega - \omega_0) + \beta_2(\omega_0)\frac{(\omega - \omega_0)^2}{2!} + \beta_3(\omega_0)\frac{(\omega - \omega_0)^3}{3!} + ... \tag{6}$$

where $\beta_k(\omega_0) = \left(\frac{\partial^k \beta}{\partial \omega^k}\right)_{\omega=\omega_0}$, ($k$ = 1, 2, 3, 4..). The term $\beta_1(=v_g^{-1})$ physically signifies the inverse of the group velocity of the pulse. The third term is the group velocity dispersion (GVD), which originates from the frequency (or wavelength) dependence of the group velocity of optical pulse in an optical medium. The GVD coefficient β_2 is related to the effective refractive index $n_{eff}(\lambda)$ as,

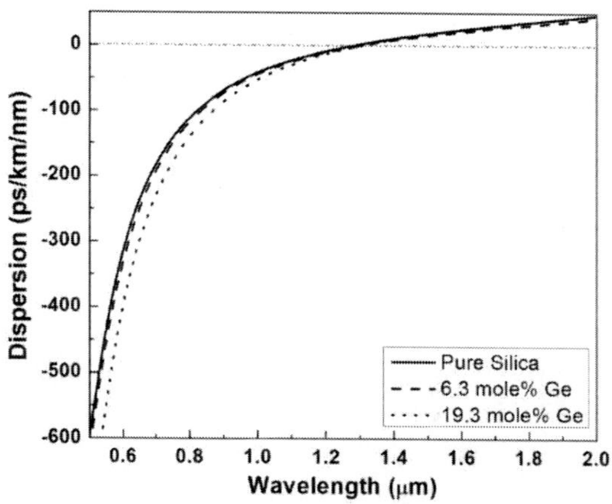

Figure 4. Typical dispersion profile of a single mode fiber (SMF) where the core is made of silica glass doped with different percentages of Germanium (Ge). The dispersion is calculated using Eqn. 5.

$$\beta_2 = \frac{\lambda^3}{2\pi c^2}\frac{d^2 n_{eff}}{d\lambda^2} \tag{7}$$

β_2 is represented in the units of ps^2km^{-1}. The GVD parameter is also represented by $D = -2\pi c\lambda^{-2}\beta_2$ in ps/km-nm. The GVD coefficient β_2 may vanish at a specific value of the wavelength λ_{ZDW} which is called the zero GVD wavelength (ZDW). We can tailor the GVD profile by adjusting the fiber geometry and manipulate the ZDW. The manipulation of ZDW is easier in specialty fiber like *photonic crystal fiber* (PCF) where we have flexibility in design. It is quite possible to have more than one ZDW depending on the nature and geometry of fiber. The sign and magnitude of β_2 are crucial for most nonlinear processes in optical fibers. A region with $\beta_2 > 0$ is called normal dispersion (ND) whereas a region with $\beta_2 < 0$ is represented as anomalous dispersion (AD). The typical dispersion profiles for single mode fiber (SMF) with different concentrations of Germanium (Ge) in the core are demonstrated in Figure 4 where the ZDW is around 1.3 μm.

Fiber Nonlinearity

When a strong electromagnetic wave interacts with a dielectric medium (like optical fiber), its response becomes nonlinear owing to the anharmonic motion of bound electrons under the influence of the strong field. Due to the nonlinear response of the medium, optical properties are changed radically yielding strong light-matter interaction resulting in intensity dependent phenomena. In this section we briefly describe the different nonlinear processes occurring in an optical fiber.

Nonlinear Processes in Silica Based Fiber

The polarisation vector **P** of a material under the influence of an external electromagnetic field **E** describes the response of the bound charge in that material. **P** maintains a linear relationship with **E** when the incident field is weak. However, for strong external field, the general relation of the polarisation vector **P** can be expressed as,

$$P = \varepsilon_0\left(\chi^{(1)}.E + \chi^{(2)} : EE + \chi^{(3)} \vdots EEE + ...\right) \tag{8}$$

Here $\chi^{(2)}$, $\chi^{(3)}$ are second and third order susceptibilities. In general the jth order susceptibility $\chi^{(j)}$ is a tensor of rank $(j + 1)$ and for isotropic or amorphous materials (such as silica-based optical fibers) it can be considered as a scalar quantity. Generally, higher order susceptibility terms are smaller and become significant only for high optical intensity. The first order susceptibility term $\chi^{(1)}$ is related to the refractive index. Due to its *centrosymmetric* structure, silica does not exhibit second order ($\chi^{(2)}$) optical nonlinearity. Hence the lowest order nonlinear effects in silica based optical fibers originate from the third order susceptibility $\chi^{(3)}$. The third order effect related to $\chi^{(3)}$ leads to interesting nonlinear phenomena like *self phase modulation* (SPM) [12], *cross phase modulation* (XPM) [13], *stimulated Raman scattering* (SRS) [14] and *four wave mixing* (FWM) [15]. The different nonlinear optical processes are briefly described below.

$\chi^{(3)}$ *Process*

The third order nonlinearity is immensely important in nonlinear fiber optics as it exhibits a wide variety of phenomena useful for versatile applications. The effects of third order nonlinearity can be demonstrated by using the superposition of the electric field **E** comprising three frequency components ω_k as:

$$E(r,t)=\sum_{k=1}^{3}E_k=\frac{1}{2}\sum_{k=1}^{3}\left[E_k^{(\omega_k)}(r,\omega_k)e^{i\omega_k t}+c.c.\right] \tag{9}$$

Exploiting the expression of the electric field **E** in Eqn. (8) one can obtain a multitude of terms with new frequencies for the third order polarization $\mathbf{P}^{(3)}$:

$$\begin{aligned}
P^{(3)} &= \frac{3}{4}\varepsilon_0\chi^{(3)}\left[\left|E^{(\omega_1)}\right|^2 E_1 + \therefore\right] && SPM \\
&+\frac{6}{4}\varepsilon_0\chi^{(3)}\left[\left(\left|E^{(\omega_2)}\right|^2 + \left|E^{(\omega_3)}\right|^2\right)E_1 + \therefore\right] && XPM \\
&+\frac{1}{4}\varepsilon_0\chi^{(3)}\left[\left(E^{(\omega_1)^2} e^{i3\omega_1 t} + c.c.\right) + \therefore\right] && THG \\
&+\frac{3}{4}\varepsilon_0\chi^{(3)}\left[\frac{1}{2}\left(E^{(\omega_1)^2} E^{(\omega_2)} e^{i(2\omega_1+\omega_2)t} + c.c.\right) + \therefore\right] && FWM \\
&+\frac{3}{4}\varepsilon_0\chi^{(3)}\left[\frac{1}{2}\left(E^{(\omega_1)^2} E^{(\omega_2)*} e^{i(2\omega_1+\omega_2)t} + c.c.\right) + \therefore\right] && FWM \\
&+\frac{6}{4}\varepsilon_0\chi^{(3)}\left[\frac{1}{2}\left(E^{(\omega_1)} E^{(\omega_2)} E^{(\omega_3)} e^{i(2\omega_1+\omega_2+\omega_3)t} + c.c.\right) + \therefore\right] && FWM \\
&+\frac{6}{4}\varepsilon_0\chi^{(3)}\left[\frac{1}{2}\left(E^{(\omega_1)} E^{(\omega_2)} E^{(\omega_3)*} e^{i(2\omega_1+\omega_2-\omega_3)t} + c.c.\right) + \therefore\right] && FWM
\end{aligned} \tag{10}$$

Here the symbol $\therefore$ stands for all possible permutation of the frequencies. The nonlinear phenomenon corresponding to each frequency term is indicated in the right hand side. The first term in Eqn. (10) corresponds to the important *self phase modulation* (SPM) effect which occurs due to an intensity dependent refractive index change.

Self Phase Modulation (SPM) and Cross Phase Modulation (XPM)

Under SPM the spectral component of the input pulse is modulated by itself and as a consequence one can have spectral broadening. The intensity dependent refractive index modulation is called the *Kerr effect*. The complex nature of $\chi^{(3)}$, however, introduces the nonlinear loss in the form of two photon absorption (TPA)[1]. In general, under *Kerr effect* the refractive index becomes intensity dependent,

$$n(\omega, I) = n(\omega) + n_2 I = n(\omega) + \bar{n}_2 |E|^2 \tag{11}$$

where $n(\omega)$ is the linear part of the refractive index, $I(=\frac{1}{2}n\varepsilon_c|E|^2)$ represents optical intensity and $\bar{n}_2$ is the nonlinear index coefficient related to $\chi^{(3)}$ as,

[1] Two photon absorption (TPA) is a nonlinear absorption process where the valence electron is excited by the absorption of two consecutive incident photons with a frequency that corresponds to an energy E=ħω above the half-bandgap energy $E_g/2$.

$\bar{n}_2 = \frac{3}{8n}\mathrm{Re}\left(\chi^3_{xxxx}\right)$. In Figure 5(a) the self-focusing effect for $n_2 > 0$ is illustrated.

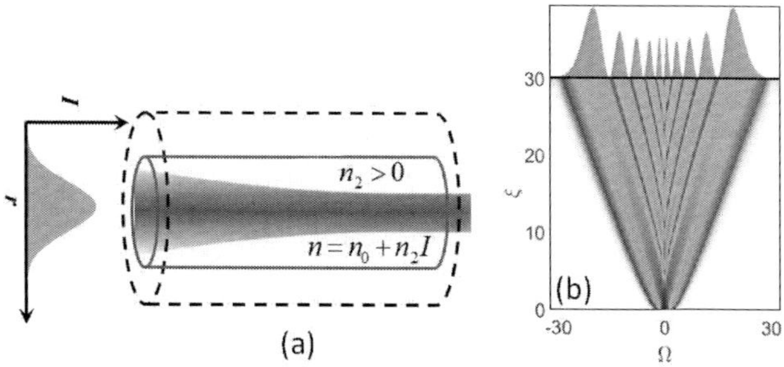

Figure 5. (a) Schematic representation of self-focusing process in optical fiber. The refractive index of the core is modified by the input intensity. (b) Spectral broadening due to SPM is illustrated where the *x*-axis corresponds to the normalised frequency (Ω) and *y*-axis represents normalised propagation distance (ξ).

The intensity dependent refractive index leads to *self phase modulation* (SPM) where an optical pulse experiences self-induced phase shift. Under SPM a propagating optical pulse experiences symmetric spectral broadening. The spectral broadening due to SPM is demonstrated in Figure 5(b). The intensity dependent nonlinear phase ϕ_{NL} accumulated due to $\bar{n}_2$ depends on propagation distance L and intensity I with $\Delta n = n_2 I$ as,

$$\phi_{NL} = \bar{n}_2 |E|^2 \times k_0 L_{eff} \tag{12}$$

where $L_{eff} = \left(1 - e^{-\alpha L}\right)/\alpha$, α being the loss coefficient.

Like SPM there is another term in Eqn. (10) called *cross phase modulation* (XPM) which is also related to the intensity dependent refractive index. In XPM the signal at frequency ω_1 is influenced by another wave of frequency ω_2. From the expression given in Eqn. (10) it is evident that the coefficient of XPM is twice that of SPM which means XPM causes modification of the refractive index that is double the modification induced by SPM. The nonlinear phase shift at ω_1 due to XPM is given by,

$$\varphi_{NL} = \bar{n}_2 k_0 L_{eff}\left(\left|E^{(\omega_1)}\right|^2 + 2\left|E^{(\omega_2)}\right|^2\right) \tag{13}$$

Unlike SPM, XPM is responsible for asymmetric spectral broadening of co-propagating optical pulses.

Raman Effect in Fiber

Under the inelastic stimulated scattering a part of optical field energy is transferred to the medium yielding the excitation of the vibrational modes of silica. In quantum mechanical picture, the annihilation of the incident photon results in the creation of a low frequency photon (*Stokes* wave) in addition to a phonon having energy that is consistent with the energy and momentum conservation. There is also a possibility to generate a high frequency photon (*anti-Stokes* wave). The formation of the phonon makes the molecule transit to an excited vibrational state. This process was discovered by the Indian scientist C.V. Raman in 1928 and is popularly known as the *Raman effect*. Practically the incident light acts as pump to generate red-shifted Stokes waves that under stimulated Raman scattering (SRS) can grow rapidly inside the medium. For a continuous wave (CW) pumping the growth of the *Stokes wave* is modeled as [16],

$$\frac{dI_s}{dz} = g_R I_p I_s \tag{14}$$

where, I_s and I_p represents the intensities of *Stokes* and *pump* wave, respectively. g_R is the Raman gain coefficient which depends on the cross section of spontaneous Raman scattering. It is customary to define $g_R(\Omega)$ as a function of $\Omega \equiv \omega_p - \omega_s$, the frequency difference between the pump and Stokes waves, to achieve the Raman gain spectrum. In general, g_R depends on the composition of the fiber core and relative polarization state of pump and Stokes waves. The Raman gain spectrum is related to the Raman response function $h_R(t)$ as,

$$g_R(\Delta\omega) = f_R k_0 n_2 \operatorname{Im}\left[\tilde{H}_R(\Delta\omega)\right] \tag{15}$$

where, $\Delta\omega = \omega - \omega_0$, $k_0 = \omega_0/c$ and $\tilde{H}_R(\Delta\omega)$ is the Fourier transform of $h_R(t)$, $\tilde{H}_R(\Delta\omega) \rightarrow h_R(t)$. The quantity f_R (= 0.18) represents the fractional contribution of the delayed Raman response to nonlinear polarization P_{NL}. Due to the amorphous nature of silica fibers it is difficult to calculate $h_R(t)$, however, assuming a single vibrational frequency Ω_R of silica molecule contributing to the Raman process, $h_R(t)$ can be analytically represented as [17],

$$h_R(t) = \left(\tau_1^{-2} + \tau_2^{-2}\right)\tau_1 e^{-t/\tau_2} \sin t/\tau_1 \tag{16}$$

where, $\tau_1 = \Omega_R^{-1}$ and τ_2 are the damping time of vibration typically having the values, $\tau_1 = 12.2$ fs and $\tau_2 = 32$ fs respectively. In Figure 6(a) and (b) the Raman response function and the Raman gain spectrum $g_R(\Omega)$ are demonstrated. Due to the amorphous nature of silica, $g_R(\Omega)$ extends over a large frequency range (upto 40 THz) with a broad peak located near 13 THz. Under a CW pump (ω_p), a signal beam whose frequency difference $\Omega = \omega_p - \omega_s$ lies within the bandwidth of the Raman gain spectrum, gets amplified [18].

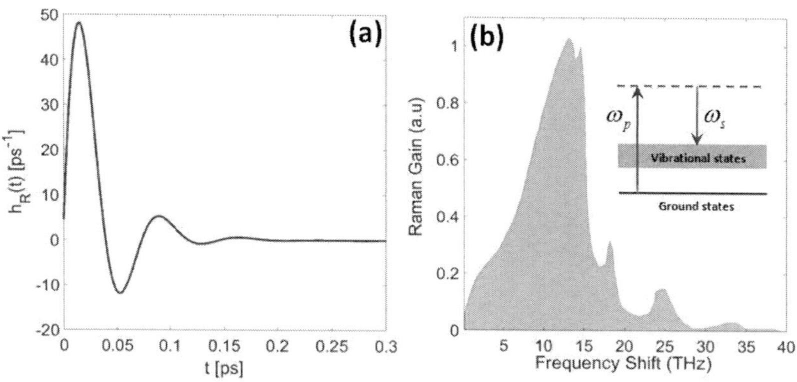

Figure 6. (a) Raman response function as a function of time. (b) Raman gain spectrum which peaks around $\Omega = \omega_p - \omega_s = 13$ THz. In the inset the Raman process in amorphous silica is schematically demonstrated.

Four Wave Mixing (FWM)

Four wave mixing (FWM) is a nonlinear interaction that occurs in the presence of multiple wavelengths in a medium, leading to the generation of new frequencies. The process can be understood as a coupling between four waves, through the real part of $\chi^{(3)}$. There are two types of FWM - in the first case three photons transfer their energy to a single photon at the frequency $\omega_4 = \omega_1 + \omega_2 + \omega_3$. This term is responsible for third harmonic generation under the condition $\omega_1 = \omega_2 = \omega_3$ which is in general difficult to achieve. In another case two photons with frequencies ω_1 and ω_2 are annihilated to create another two photons with frequencies ω_3 and ω_4 in a manner that energy conservation is maintained through the following equation,

$$\omega_1 + \omega_2 = \omega_3 + \omega_4 \tag{17}$$

In addition, due to the momentum conservation between the four interacting photons, the phase mismatch between the pump waves must be zero. This can be expressed by,

$$\Delta k = \beta(\omega_1) + \beta(\omega_2) - \beta(\omega_3) - \beta(\omega_4) + 2\gamma P_0 = 0 \tag{18}$$

where the term $2\gamma P_0$ accounts for the nonlinear phase shift of the intensity dependent refractive index. In the most general non-degenerate case, , $\omega_1 \neq \omega_2$, one must launch two pump beams in order to excite the FWM process. However a special case may also arise when $\omega_1 = \omega_2$ (degenerate case) and for this condition FWM can be initiated with a single pump beam. Physically two sidebands are generated symmetrically at frequencies ω_3 and ω_4 with frequency shift,

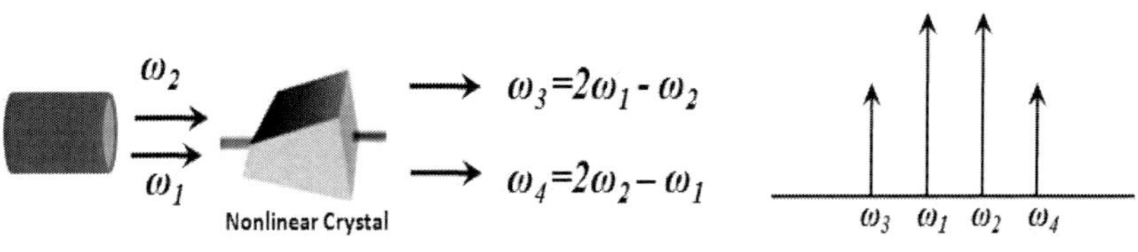

Figure 7. Schematic diagram showing FWM in a nonlinear crystal.

$$\Delta\omega = \omega_1 - \omega_3 = \omega_4 - \omega_1 \tag{19}$$

where it is assumed that $\omega_4 > \omega_1 > \omega_3$, *i.e* the lower frequency sideband is at ω_3 and the higher frequency side band is at ω_4. The lower and higher frequency sidebands are referred to as Stokes and anti-Stokes bands respectively. In Figure 7 the mechanism of FWM is illustrated graphically.

Soliton Formation

The interplay between dispersive and nonlinear effects leads to optical soliton which is a special kind of wave packet that can propagate undistorted over a long distance [16]. In normal dispersion domain the red shifted part of the pulse spectrum propagates faster than the blue shifted part resulting in pulse broadening in time domain. Like SPM, dispersion also introduces a phase component as, $\phi_D(\xi,\tau) = -\mathrm{sgn}(\beta_2)\frac{\xi\tau^2}{2(1+\xi^2)} + \frac{1}{2}\tan^{-1}(\xi)$. The instantaneous frequency $\omega_D(\tau)\left[= \mathrm{sgn}(\beta_2)\frac{\xi}{1+\xi^2}\tau\right]$ becomes a linear function of time such that for anomalous dispersion ($\beta_2 < 0$), $\omega_D(\tau)$ becomes positive at the leading edge ($\tau < 0$) and negative at the trailing edge ($\tau > 0$), which is opposite to that of SPM. With suitable choice of parameters, the chirp induced by SPM and dispersion can be counterbalanced and we obtain an optical soliton. The most standard equation that describes the solitonic pulse propagation is given as,

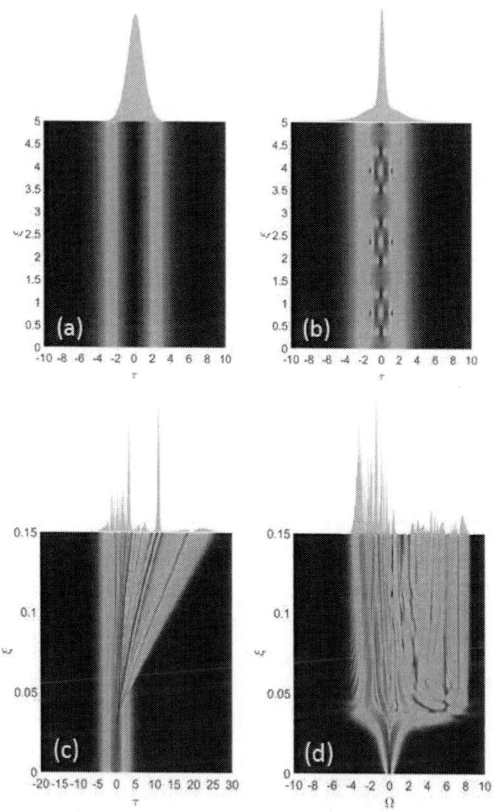

Figure 8. Evolution of fundamental (N=1) (Figure a) and higher order (N=3) (Figure b) soliton simulated using eqn. 20. Pulse dynamics during SC generation in (Figure c) time and (Figure d) frequency domain simulated using eqn. 21.

$$i\frac{\partial\psi}{\partial\xi}+\frac{1}{2}\frac{\partial^2\psi}{\partial\tau^2}+N^2|\psi|^2\psi=0 \tag{20}$$

For $N = 1$, an input pulse with the amplitude $\psi(0,\tau) = \text{sech}(\tau)$ propagates as the fundamental soliton with the general solution $\psi(\xi,\tau) = \text{sech}(\tau)e^{i\xi/2}$. In Figure 8(a) and (b) the evolution of fundamental ($N = 1$) and higher order soliton ($N = 3$) is demonstrated.

Extreme Nonlinear Process: Supercontinuum Generation

Supercontinuum (SC) generation is an optical phenomenon where a collection of nonlinear processes (like SPM, Raman scattering, shock effect etc) act together upon an input pump beam to cause severe spectral broadening at the output [19]. The interplay between nonlinearity and dispersion is crucial for SC generation. Highly nonlinear photonic crystal fiber (PCF) becomes an excellent waveguide to excite the SC as it offers tight confinement of the optical field which effectively enhances the nonlinear parameter $\gamma(=n_2k_0/A_{eff})$. The SC process can be modeled by the following generalised nonlinear Schrödinger equation,

$$\frac{\partial A(z,t)}{\partial z}=i\sum_{m=2}^{\infty}i^m\frac{\beta_m}{m!}\frac{\partial^m A}{\partial t^m}+i\gamma\left(1+\frac{i}{\omega_0}\frac{\partial}{\partial t}\right)A(z,t) \times\int_{-\infty}^{t}R(t-t')|A(z,t)|^2\,dt' \tag{21}$$

where $R(t\text{-}t')$, $\beta_m\left(=\left.\frac{\partial^m\beta}{\partial\omega^m}\right|_{\omega=\omega_0}\right)$ are the Raman response function and mth order dispersion coefficient respectively. The temporal and spectral dynamics of SC is demonstrated in Figure 8(c) and (d), respectively. In soliton based SC generation the intrapulse Raman scattering (IPRS) plays a very dominant role in generating low frequency spectral components (*red* components). For pulse width of the order of few hundreds of fs having spectral width large enough, the Raman gain can amplify the low-frequency components of a pulse, with the high-frequency components of the same pulse acting as a pump. This process continues during pulse propagation resulting in spectral red-shift. The phase matched radiation is responsible for the formation of high frequency component (*blue* component) during SC process. When an optical soliton is perturbed by third order dispersion it emits dispersive wave at specific frequency which corresponds to the *blue* component [16].

Fundamentals of Photonic Crystal

A photonic crystal (PC) is an optical material having a periodic dielectric profile which can prevent light of certain frequencies or wavelengths from propagating in one, two or any number of polarization directions within the material [20]. One of the most important features of photonic crystals is the presence of a photonic bandgap (PBG). Light with wavelengths that lie in the photonic bandgap cannot exist in the crystal, giving rise to completely different optical phenomena. A quarter-wavelength plate or multiple-layer dielectric mirrors such as those used for a high-Q-value laser cavity are simple examples of one-dimensional (1D) *photonic crystals*. However in their seminal work, E. Yablonovitch [1] and S. John [2] theoretically demonstrated that more effective control of light can be possible with 2D and 3D PCs. E. Yablonovitch demonstrated the existence of a *photonic band gap* (PBG) where all modes are missing in all directions, and consequently, spontaneous emission corresponding to the gap energy is inhibited inside a 3D PC. On the other hand, S. John pointed out that new phenomena can be observed by using a PC, which includes localization of light.

Types of Photonic Crystals

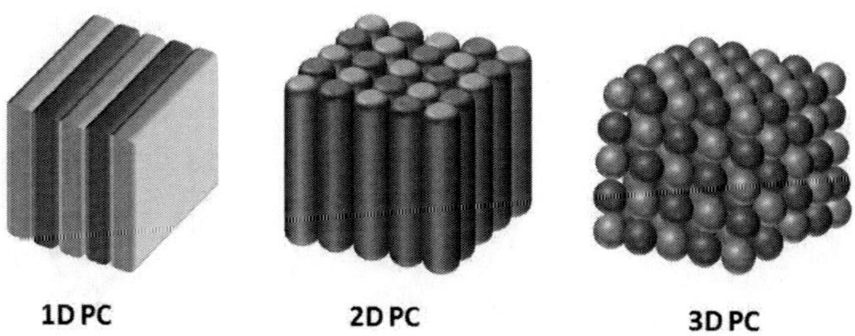

Figure 9. Schematic representation of 1D, 2D and 3D PC.

A PC is defined as a crystal whose dielectric constant (ϵ) varies periodically in specific directions. In the case where the variation is along one direction, we call it a 1D PC. Accordingly, we define 2D and 3D PCs corresponding to the respective cases where ϵ varies along two and three independent directions. In Figure 9 we schematically demonstrate the structures of 1D, 2D and 3D PCs where the different colors represent materials with different dielectric constants. The 1D PC can simply be realized by the multilayer dielectric film. In the case of a 2D PC as shown in Figure 9, rods of two different materials form a 2D photonic crystal plane. As an extension, the 3D PC is formed by the alternative arrangement of two different materials. In 1D PC, the eigenmodes of light with wave vector **k** parallel to the direction of periodicity are quite different from a uniform dielectric. Similarly, the respective eigenmodes of light with **k** in the 2D plane of a 2D PC and with **k** in an arbitrary direction in the 3D PC become unique as compared to homogeneous 2D and 3D dielectrics.

1D Photonic Crystals

The most general example of the 1D PC structure is the multilayer stack consisting of two alternating layers of different materials, with refractive indices say n_1 and n_2, and period say Λ. Such structures transmit electromagnetic waves at certain frequencies while reflecting others. During the wave propagation under normal incidence, each layer of the 1D periodic structure reflects a fraction of wave. For specific frequencies the optical path differences of the reflected wavelets, along with phase changes introduced at the layer boundaries, cause the reflecting wavelets to interfere constructively. These frequencies are eventually prevented from transmission and lie in the regime of photonic band-gap.

2D Photonic Crystals

An ideal 2D PC yields a periodic dielectric variation in two directions and is homogeneous in the third. A common example of 2D PC is the square or hexagonal lattice of dielectric columns or the air hole structure in the dielectric. In case of a 2D PC when the electric field **E** (magnetic field **H**) is perpendicular to the wave vector **k** it is called TM (TE) mode. The photonic band structure for the square and hexagonal lattice is highly sensitive to the polarization state of the light. For example, in square lattice we have the photonic band gap for TM modes but there is no band gap for TE mode. However the hexagonal lattice offers photonic band gaps for both TE and TM polarizations. It is interesting to note that a defect in such structure can *trap* or *guide* light [20]. While point defects can trap light, linear defects can guide light from one location to another and act as photonic channel waveguides.

3D Photonic Crystals

In a 3D photonic crystal, the dielectric constant is tailor-made to vary periodically (in the order of the wavelength of light) in three dimensions, giving rise to an omnidirectional photonic band gap (PBG) which prohibits electromagnetic propagation in all three directions and substantially modifies the dispersion around a specific frequency region. 3D PCs have become an interesting topic as they reveal exciting phenomena such as the suppression and/or enhancement of light emission useful for unique device formation. For a complete PBG at optical wavelengths, a high-resolution 3D fabrication technique and high-dielectric constant materials are required. 3D PC offers a strong light-matter interaction but at the same time fabrication of such material is a challenge.

State of Light in Vacuum vs. Photonic Crystal

In case of a uniform and transparent medium the eigenstate (mode) of light is denoted by a set of indices ($\hbar\omega$, $\mathbf{k}$, $\mathbf{e}$) where $\hbar\omega$ and $\mathbf{k}$ refer to the photon energy and wave vector respectively and the polarization state is defined by $\mathbf{e}$. Note, for each eigenstate a characteristic spatial pattern of electric (**E**) and magnetic (**H**) field is specified, which is called the *mode*. On the other hand, for each polarization, there is a dispersion relation between $\hbar\omega$ and $\mathbf{k}$. For example, in a sample of infinitely large space of vacuum, the relation is expressed as $\omega = ck$ ($k = |\mathbf{k}|$) where c refers to the velocity of light. Here the eigenstate of a transverse plane wave of k with an arbitrary magnitude and direction is allowed to exist. Next, as an example of the characteristic modes of electromagnetic waves in a closed space (vacuum), let us consider the case of a cube of side l in each of the x, y and z directions. By adopting the periodic conditions on the surfaces of the cube of side l, we have $k_{j=x,y,z} = (2\pi/l)n_j$ and $n_j = 0, \pm1, \pm2, \pm3, \ldots$ Therefore, the modes with a set of discrete values of $(2\pi/l)(n_x, n_y, n_z)$ for (k_x, k_y, k_z), are allowed with k expressed as $k^2 = (2\pi/l)^2(n_x^2 + n_y^2 + n_z^2)$. Apart from the freedom of 2 types of polarization, the number of each point (n_x, n_y, n_z) in the sphere of radius R corresponds to the number of modes with frequencies in a range from 0 to $\omega = (2\pi c/l)R$, where $R =$

$(n_x^2+n_y^2+n_z^2)^{1/2}$. The number of modes in the volume segment $R \rightarrow R + dR$ determines the number of modes in the frequency range $\omega \rightarrow \omega + d\omega$ as $2\times 4\pi R^2 dR \rightarrow \frac{\omega^2}{\pi^2 c^3} l^3 d\omega$. Hence the density of state of the photon $\rho(\omega)$ i.e., the number of modes per unit volume with frequencies between $\omega \rightarrow \omega + d\omega$ is given as, $\rho(\omega)d\omega = [\omega^2/(\pi^2 c^3)]d\omega$. Note, the mode density is completely different in case of PC compared to the free space [20].

The eigenstate of the light in PC varies drastically compared to the vacuum where susceptibility $\epsilon(\mathbf{r})$ varies periodically. For non-magnetic dielectric the wave equation of the magnetic field $\mathbf{H}(\mathbf{r})$ can be expressed as,

$$\nabla \times \left[\varepsilon\ r^{-1} \nabla \times H\ r\right] = \omega / c^{\,2}\ H\ r \tag{22}$$

which is an eigenvalue equation with hermitian operator $\nabla\times\varepsilon(\mathbf{r})^{-1}\nabla\times$ having real eigenvalue $(\omega/c)^2$. For a given PC, $\epsilon(\mathbf{r})$, one can solve the master equation Eqn. (22) to get the eigen modes. The electric field $\mathbf{E}(\mathbf{r})$ can be obtained from $\mathbf{H}(\mathbf{r})$ by using the relation,

$$\mathbf{E}\ \mathbf{r} = i\omega^{-1} c\left[\varepsilon\ \mathbf{r}^{-1} \boldsymbol{\nabla}\times\mathbf{H}\ \mathbf{r}\right] \tag{23}$$

Solving Eq. (22) for a given $\epsilon(\mathbf{r})$ is the key problem in calculating photonic band gap. Generally the eigen energy $\hbar\omega$ is obtained as a function of wave vector $\mathbf{k}$ and depicted in the reduced Brillouin zone. In PCs the dispersion curve of ω vs $\mathbf{k}$ breaks to form photonic bands (PBs) as represented in Figure 10.

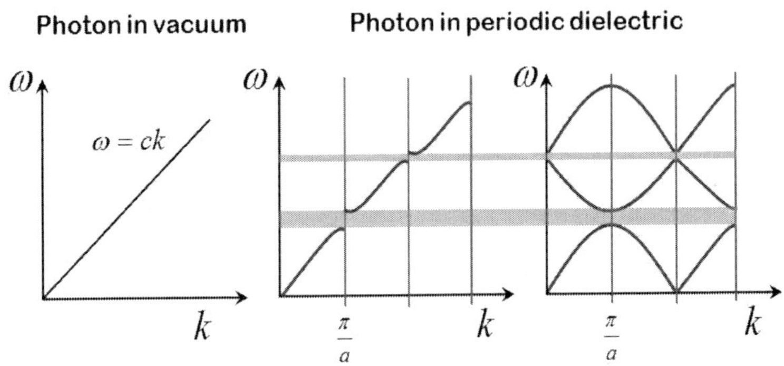

Figure 10. Linear dispersion relation when a photon moves through vacuum. The dispersion relation breaks at the Brillouin zone boundary represented by $k = \pi/a$ yielding PBG.

Physical Origin of Photonic Band Gap

A photonic crystal has a periodic dielectric constant mathematically defined as, $\epsilon(\mathbf{r}) = \epsilon(\mathbf{r} + \mathbf{R}_i)$, where $\mathbf{R}_i$ represents the primitive lattice vectors. According to the *Bloch* theorem, Eqn. (22) should have the solution of the form $\mathbf{H}_n(\mathbf{r}) = \psi_{\mathbf{k}}(\mathbf{r})e^{i\mathbf{k}\cdot\mathbf{r}}$, where $\psi_{\mathbf{k}}(\mathbf{r} + \mathbf{R}_i) = \psi_{\mathbf{k}}(\mathbf{r})$, with eigenvalues $\omega_n(\mathbf{k})$ satisfying $\omega_n(\mathbf{k}+\mathbf{G}_j) = \omega_n(\mathbf{k})$, where $\mathbf{G}_j$ is the primitive reciprocal lattice defined by $\mathbf{R}_i \cdot \mathbf{G}_j = 2\pi\delta_{ij}$. The eigenvalues $\omega_n(\mathbf{k})$ come out to be periodic in PCs and we can plot $\omega_n(\mathbf{k})$ as a function of $\mathbf{k}$ in the first *Brillouin zone* (BZ) to get the band diagram. In simple 1D system with uniform dielectric constant ϵ the modes lie along the *light line* $\omega(k) = ck/\epsilon$ (see Figure 10 (left image)). For such mode, there is a degeneracy in the folded BZ boundary $k = \pi/a$. Now if the dielectric constant ϵ is perturbed with a periodicity a, the degeneracy of the mode is removed as the two standing modes are distributed accordingly. The standing modes are distributed in two possible ways, the nodes of the modes are placed either in the center of low ϵ layer or high ϵ layer. Any other distribution of the modes in the periodic layer is prohibited as it violates the symmetry. Due to such specified distribution of the modes in the perturbed system, the energy concentration is different in different dielectric layers which breaks the degeneracy and a band gap at 1st BZ boundary is formed as demonstrated in Figure 10.

Light Guidance in Fiber with Photonic Crystal

In an infinite photonic crystal, the guidance of electromagnetic wave is prohibited for the frequencies that lie within a total photonic band gap. However localized defect states for isolated frequencies within the band gap can emerge if a defect is introduced in the infinite lattice analogous to the bound states associated with defects in semiconductors.

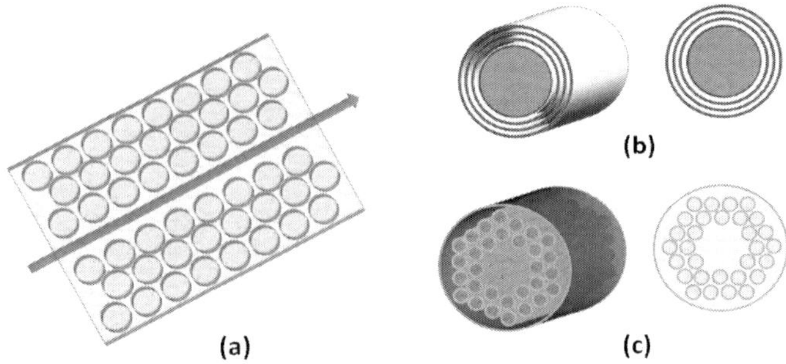

Figure 11. Photonic crystal waveguides. (a) A 2D PC waveguide where the light can be guided through the defect channel as indicated by arrow. (b) *Bragg fiber*, a photonic structure, where multiple concentric dielectric layers create the PBG which restricts the transverse flow of light. (c) Cylindrical structure with periodic air hole running through the length. The periodic air hole structure yields the PBG and the light can propagate through the central defect.

For a 3D PC having single point defect the light can be confined in the vicinity of the defect, whereas for a linear defect (in the form of a channel in 2D PC) the light can propagate along it (see Figure 11(a)). The periodic photonic structure in the transverse direction acts as a perfect mirror for a certain band of frequencies and the light can be guided along the perpendicular direction like conventional optical fiber. Such waveguide can be realised with a cylindrical structure having successive layers of different dielectric material (see Figure 11(b)) or alternatively by drilling periodic holes all the way in a cylindrical structure having a defect in the form of core (see Figure 11(c)). The light is trapped in the core by an out-of-plane photonic band gap (PBG) created by the 2D PC cladding. It is interesting to note that while exploiting a PBG material for the cladding, reflection at some frequency is guaranteed regardless of the refractive index of the material inside the defect. A defect in a PC can hence confine and guide light in low refractive index media, such as a gas (air for example) or vacuum.

Photonic Crystal Fiber

The development of optical fiber technology has progressed at an incredible rate from the demonstration of the first ever low loss single mode fiber in the 1970s to being the key component of the present day sophisticated global telecommunication network. With time, optical fibers, with suitable modification in their design and betterment in performance, found applications in domains other than telecommunication. Specialty optical fibers are now in huge demand in industrial, healthcare, strategic sectors and others. Since early 1996, a new type of specialty optical fiber known as photonic crystal fiber (PCF) has been in practical existence alongside all-solid specialty optical fibers. PCFs contain a periodic transverse microstructure and are mainly of two types – (a) Solid core PCF which is better known as Microstructured Optical Fiber (MOF) and (b) Hollow core PCF (HC-PCF). HC-PCF can be further divided into two main sub-groups: Photonic Band Gap (PBG) fiber and Kagomé fiber. The cross sections of the solid core PCF, PBG fiber and Kagomé fiber are shown in Figure 12. Three distinct types of guidance mechanisms exist in PCFs depending on their core and microstructure,

i. Modified total internal reflection in solid core PCFs.
ii. Photonic band gap guidance in hollow core Photonic Band Gap (PBG) fibers.
iii. Inhibited coupling between core and cladding modes in hollow core Kagomé fibers.

Right from its origin, PCFs have stirred excitement among the research fraternity since they offer flexibility in tailoring their geometrical parameters, namely air hole diameter (d) and pitch (Λ - distance between the centres of two adjacent holes). The fabrication techniques have become more sophisticated to accommodate the manufacture of microstructures to accuracies of 10 nm on the scale of 1 μm. This enables remarkable control of key optical properties like birefringence, single mode guidance, dispersion, nonlinearity etc. Thus, PCFs have extended the range of possibilities in optical fibers in two ways - by improving the already established properties and introducing new features unachievable in conventional fibers.

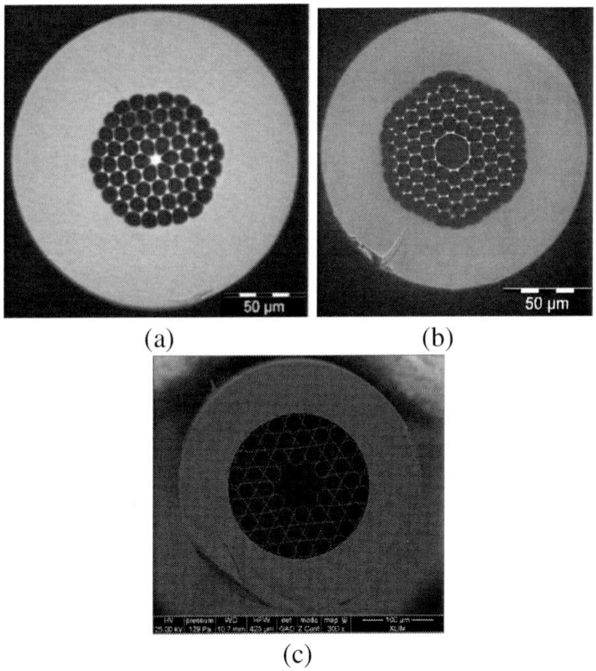

Figure 12. Cross-section of a (a) Solid core PCF and (b) Hollow core PCF fabricated at CSIR-CGCRI, Kolkata. (c) Cross-section of a Kagomé fiber. The fiber was fabricated at XLIM, University of Limoges, France.

Light Guidance Mechanisms in PCF

Modified Total Internal Reflection

In solid core PCFs the cladding is made up of a hexagonal array of air holes embedded in silica matrix that reduces the effective refractive index of the microstructured cladding from that of the solid silica core. Light guidance through the core resembles the conventional total internal reflection but it has some unique features that markedly distinguish it from step index fibers and hence it is called modified total internal reflection. The unique features, which will be studied later in subsequent sections, arise due to the piecewise discontinuous nature of the core boundary where sections of air holes strongly blocking the light are interspersed with sections of silica glass that aid leakage of light [21].

Photonic Band Gap (section and sub-section captions may be included in the same page as the text)

The idea of trapping light in a hollow core of an optical fiber by creating a 2-D photonic crystal cladding with the help of microscopic air holes propagating along the entire length of the fiber was first proposed by Prof. Philip Russell [21]. The 2-D photonic crystal cladding, when appropriately designed, would form an out-of-plane photonic band gap (PBG) which consequently would prevent the light incident from air into the core from escaping into the cladding. Such a guidance mechanism is unique in the sense that it avoids the need for TIR and allows light propagation through low-index hollow core. The term out-of-plane refers to the presence of a PBG for a non-zero propagation constant, β along the direction of the fiber axis such that no photonic state i.e., optical mode with **E** and **H** field dependencies exist over a given range of wavelength [22]. The propagation regimes of a PBG fiber can be properly understood by considering the schematic diagram shown in Figure 13. The three dispersion lines - air light line, photonic crystal (PC) line and silica line mark the boundary between allowed and forbidden propagation in a particular medium. For the air (or vacuum) light line, $k = \beta$. Propagation in air is forbidden below this line for the region $k < \beta$. Since silica is the material with the highest index in a PCF so no photonic state is supported below the silica line given by $n_{Si}k = \beta$ and this line marks the cut-off region. The intermediate PC line given by $n_{FSM}k = \beta$ represents the cut-off for the photonic crystal cladding structure below which no cladding state can exist. The finger-like structures in Figure 13 represent the PBGs that can be seen extending into the region in which propagation in air is supported. At points within these fingers above the air light line, light is free to propagate in air which makes band gap guidance in a hollow core possible.

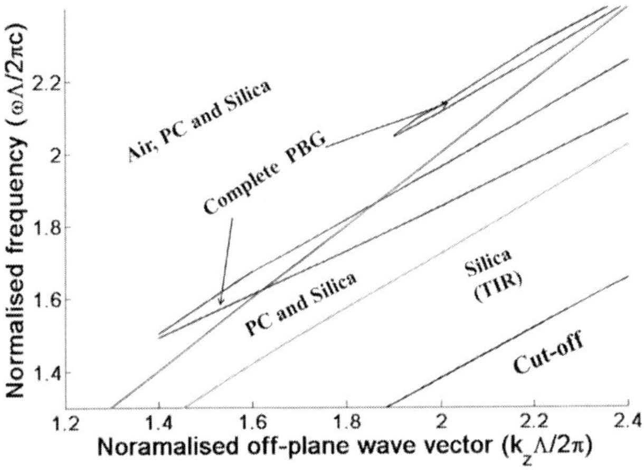

x-axis title to be corrected – it should be 'Normalised' and not 'Noramalised'

Figure 13. Propagation diagram for a hexagonal array of air holes in silica glass. The different regions where light is cut-off or able to propagate are marked. The fingers represent the positions of full 2D photonic band gap. Light can be guided in air at positions within these fingers above the air light line (Courtesy: Dr. Tushar Biswas).

Broadband Hollow Core Guidance – Inhibited Coupling

PBG fibers are limited by the bandwidth for guidance which depends on the achievable width of the bandgap close to the air light line shown in Figure 13. This is because the presence of the glass struts connecting the glass nodes restricts the frequency range of the broadest bandgap in practical air-glass structure to about 15% of the central frequency of the bandgap. Although recently developed bandgap fibers with high air filling fractions open up higher order bandgaps but these are narrow and separated by broad spectral regions with high attenuation loss. Hence the Kagomé fiber, which provides broad spectral range of relatively low attenuation, was developed to overcome the limitation of narrow band guidance. In such a fiber, the cladding does not create a band gap but is designed in a way that prevents the cladding modes close to the air/vacuum light line from interacting strongly with the core mode i.e., inhibit the coupling between the modes present in the glass nodes surrounding the hollow core and the core mode itself. In contrast to PBG fibers, the glass nodes in Kagomé fibers are required to be as small as possible as otherwise there remains a possibility of resonance close to the air light line at wavelengths distinct from the transverse resonances associated with the struts, thus narrowing the guiding bandwidth available. Another structural difference between PBG fiber and Kagomé fiber is the pitch. While in PBG fibers the wavelength range of operation is determined by the pitch, in the case of Kagomé fibers it is the strut thickness that determines the guidance range.

Properties of PCF

Endlessly Single Mode

One of the most striking features of a PCF is that, when appropriately designed, it can remain single mode even at very short wavelengths and hence demonstrate endlessly single mode (ESM) configuration. The endlessly single mode behavior of PCFs was first explained by Birks et al. [23]. In a standard step index fiber having core radius a and core and cladding index n_{co} and n_{cl} respectively, the number of guided modes is given by the V -parameter:

$$V = \frac{2\pi a}{\lambda}\sqrt{n_{co}^{\ 2} - n_{cl}^{\ 2}} \tag{24}$$

For single mode guidance, the V-parameter should be less than 2.405. Thus, at sufficiently short wavelengths, standard fibers are multimode. The expression of V-parameter for a PCF gets modified as follows:

$$V = \frac{2\pi a_{eff}}{\lambda}\sqrt{n_{co}^{\ 2} - n_{FSM}^{\ 2}} \tag{25}$$

where a_{eff} is the core radius and n_{FSM} is the effective cladding index. In PCFs, the photonic crystal cladding experiences strong wavelength dispersion. The field stays concentrated in the solid core and avoids the holes at shorter wavelengths thus increasing the effective cladding index, n_{FSM}. This dispersion counteracts the inverse dependence of *V*-parameter on wavelength and helps to maintain the *V* -parameter value below 2.405 even at shorter wavelengths by appropriately choosing the d/Λ value. It was later shown numerically that a PCF exhibits ESM behaviour when its $d/\Lambda \leq 0.43$.

Alternatively, the endlessly single mode behavior in PCF structures can be understood by viewing the microstructured cladding as a modal filter or *sieve*. The transverse effective wavelength, given by $\lambda_{eff}^{g} = 2\pi\ (k^2 n_g^{\ 2} - \beta^2)^{-1/2}$, of the fundamental mode in the solid core is nearly

4Λ which does not allow the mode to squeeze through the silica struts between the holes since the width of the silica channels are below the *Rayleigh* resolution limit. However, for relatively smaller d/Λ value, the higher order modes are able to escape because their transverse effective wavelengths are shorter and hence resolution is higher.

Dispersion

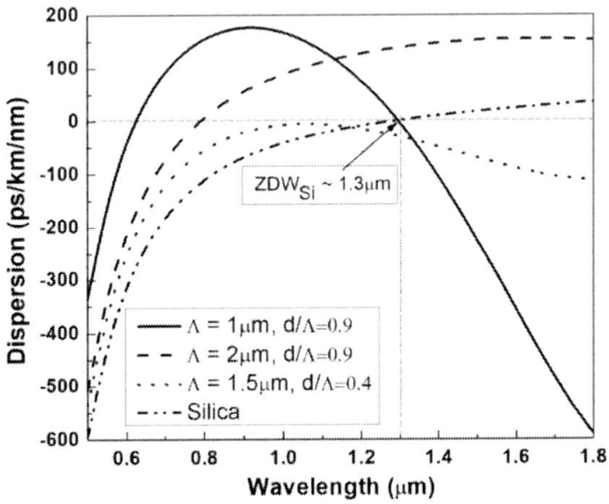

Figure 14. Dispersion profiles of solid core PCFs with different geometrical parameters showing the tunability of the ZDW. The PCF configuration represented by the dotted curve shows normal dispersion for the complete spectral window, the dashed curve represents a PCF structure that has a single ZDW but shifted towards shorter wavelength while the PCF structure represented by the solid curve demonstrates two ZDWs, one of which is shifted to the visible domain. The dash-dotted curve shows the dispersion profile of pure silica having ZDW at $\approx 1.3\mu m$.

The flexibility in tailoring the design parameters of PCFs also leads to novel dispersion properties that are unachievable in standard fibers and can be tailored and controlled with unprecedented freedom. As mentioned before, the large difference between the core and cladding refractive indices and the ease of modifying the hole size and arrangement makes the cladding index strongly dependent on wavelength leading to unusual dispersion properties in PCFs. PCFs offer greatly enhanced control of the magnitude and sign of the dispersion as a function of wavelength which represents an even greater opportunity than a mere enhancement of the effective nonlinear coefficient.

Basically, altering the geometrical parameters significantly modifies the waveguide dispersion which is manifested as unusual positions of the zero dispersion wavelength (ZDW) that can be tuned over a wide range e.g., in the visible spectral regime [24, 25], multiple ZDWs, ultraflattened dispersion slope over a long wavelength range [26, 27] are some such interesting properties that can be achieved by suitably designing a PCF. Figure 14 shows how dispersion can be controlled by tuning the geometrical parameters of a PCF.

Nonlinearity

Besides the above-mentioned properties, the easily tailorable geometrical parameters of PCFs also lead to tuning the fiber nonlinearity over a wide range. The nonlinear coefficient of an optical fiber is given by,

$$\gamma = \frac{2\pi n_2}{\lambda A_{eff}} \tag{26}$$

where, λ is the operating wavelength and n_2 is the nonlinear index coefficient of the core material and is related to the third order susceptibility. Typically, $n_2 = 2.7\times10^{-20}\ m^2/W$ at the wavelength of 1.06 μm for silica. A_{eff} is the effective mode area expressed as [16],

$$A_{eff} = \frac{\left[\int_{-\infty}^{\infty}\int_{-\infty}^{\infty} |E\ x,y|^2\, dxdy\right]^2}{\int_{-\infty}^{\infty}\int_{-\infty}^{\infty} |E\ x,y|^4\, dxdy} \tag{27}$$

where, $E(x, y)$ is the electric field. The mode area of a PCF can be tuned to a large extent by appropriately modifying the hole diameter and pitch. A_{eff} can be largely increased by increasing the pitch and reducing the hole diameter. Large mode area allow efficient transmission of high power light without any distortion before the onset of intensity-related nonlinearities or damage and hence is beneficial for delivery of high laser power, fiber amplifiers and fiber lasers. Alternatively, the mode area in PCF can be reduced by decreasing the pitch and increasing the hole diameter. This helps in tight mode confinement in a very small area that gives rise to enhanced nonlinearity in a PCF. In complete contrast to the solid core nonlinear PCF, HC-PCFs demonstrate extremely low level of nonlinearity because of light propagation in air that significantly reduces the overlap between glass and light [21]. The nonlinearity in PCF can be controlled upto the fourth order of magnitude when appropriately designed.

Birefringence

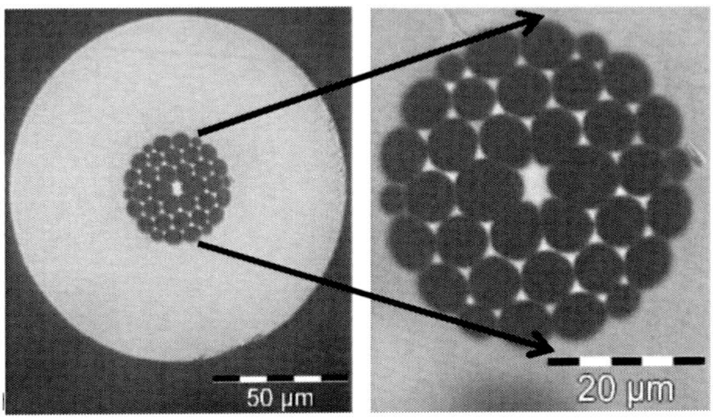

Figure 15. Cross-section of a birefringent solid core PCF fabricated at CSIR-CGCRI where it can be seen that the core is made elliptical by changing the diameter of two opposite holes surrounding the solid core.

In general, a circularly symmetric core and cladding structure in a fiber does not offer birefringence since the modes are degenerate. However, in PCFs, the freedom of tuning the hole diameter allows us to deliberately break the six-fold symmetry of the circular core and develop two-fold symmetry that induces birefringence. PCFs can potentially offer ten times higher birefringence than conventional fibers [28] by changing the diameter of two opposite air holes around the core that makes the core elliptic. This can be achieved by using capillaries of different wall thicknesses around the solid core during fabrication by the well-known stack and draw method. The cross section of one such birefringent solid core PCF fabricated at CSIR-CGCRI, Kolkata is shown in Figure 15. An added advantage is that the birefringence in PCF is almost a hundred times less sensitive to temperature compared to conventional polarization-maintaining fibers [29] which are fabricated using at least two different glasses that have different thermal coefficients and hence generate temperature-dependent stresses.

Applications of PCFs

Supercontinuum Generation Using Solid Core PCF

The general process of supercontinuum (SC) generation has already been defined and explained in a previous section. Although SC generation has been investigated in a wide variety of nonlinear media and various types of waveguides because of its applications in diverse fields, it was the advent of PCFs that revolutionized the research on SC generation. This is because of the design flexibility that PCFs offer in the form of endlessly single mode guidance, easily tailorable dispersion and high nonlinearity due to tight modal confinement. All these properties of PCFs allow SC generation to be observed over a much broader range of source parameters than has been possible with bulk media or conventional fibers. In PCFs, SC can be generated using even high power continuous wave (CW) sources besides unamplified input pulses from tens of femtoseconds to several nanoseconds. Nonlinear processes like Self Phase Modulation (SPM), Cross Phase Modulation (XPM), Four Wave Mixing (FWM), Modulation Instability (MI), Intrapulse Raman Scattering (IPRS) and soliton dynamics all play important roles along with dispersion of the fiber in generating the ultra-broadband spectra. The input pulse width and the location of the zero dispersion wavelength (ZDW) mainly determines which nonlinear processes will be predominant in generating the SC. For example, for a PCF pumped in the anomalous dispersion domain with a femtosecond pulse, soliton dynamics and IPRS play pivotal roles in generating the broadband spectrum whereas for normal domain pumping with picosecond/nanosecond pulses, SPM, MI and FWM participate in SC generation. Figure 16 (b) shows the evolution of the SC spectrum in one of the CSIR-CGCRI fabricated solid core PCFs. SC generation can be numerically simulated by solving the Generalized Nonlinear Schrödinger Equation (GNLSE) given by Eq. (21) and already described in earlier section. Here, we have used the Split Step Fourier Method to solve the GNLSE in order to compare, correlate and analyze the experimental and theoretical data.

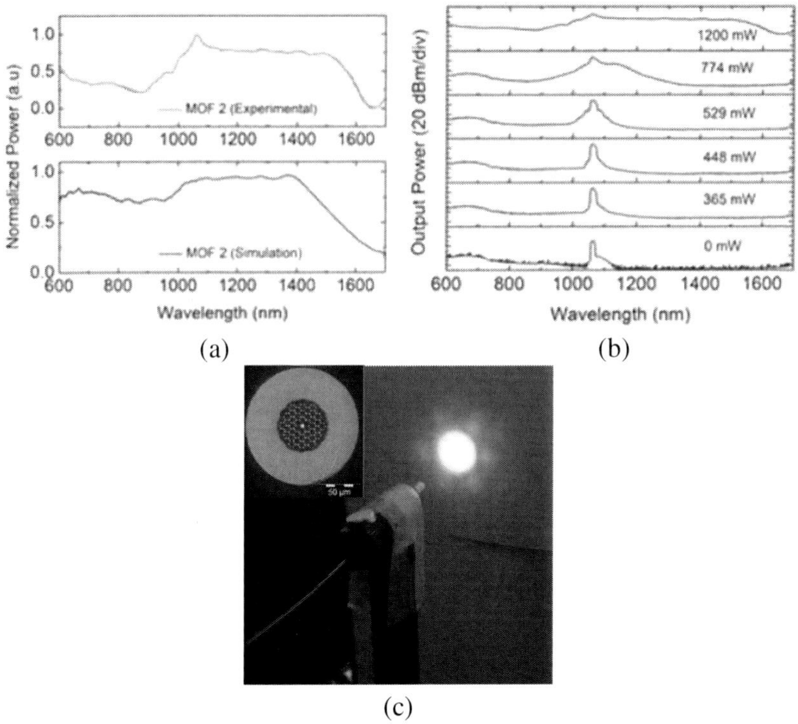

Figure 16. (a) Supercontinuum generation in an in-house fabricated solid core PCF (the cross-section is shown in the inset of (c)) using fs pulses at pump wavelength of 1064 nm. (b) Evolution of SC as a function of input power. (c) White-light generation during SC process. The experiment is done at CSIR-CGCRI, Kolkata.

Gas-Filled Hollow Core PCF

HC-PCFs offer the advantage of filling the core and/or the cladding with different gases and liquids thereby opening up exciting prospects for applications of gas-laser interaction in various fields like developing laser sources at unconventional wavelengths, nonlinear optics for supercontinuum generation beyond the transmission window of silica, spectroscopy [30], frequency metrology, quantum/coherent optics and so on. Among them, ultrafast and nonlinear applications are the most studied. Mostly inert gases like Helium (He), Argon (Ar), Krypton (Kr) and Xenon (Xe) are used to pressure-fill the hollow core. The Kagomé fibers are suitable candidates for laser-gas phase study since they offer broadband guidance. Additionally, the dispersion of the fiber can be tailored by tuning the pressure of the gas filled in the core which gives rise to the possibility of shifting the ZDW across the deep UV spectral region.

Plasmonic Sensor

An interesting formalism has been introduced in HC-PCF where basic principle of surface plasmon polariton in composite dielectric is exploited. When the cladding holes of HC-PCF are filled selectively with pure metal (Figure 17 (a) – hole on the RHS of hollow core) it would find potential application in bio-fluid sensors where the hollow core is used as an integrated microfluidic channel to pass bio-analytes. It is observed that a minute change in RI value of the analyte causes shifting of the surface plasmon resonance (SPR) wavelength resulting in loss of intensity of the core-guided mode (Figure 17 (b)) [31]. Unlike planar metal-dielectric interface which produces only one SPR mode, the cylindrical surface of HC-PCF can generate several SPRs due to various possible SPP modes (Figure 17 (c)). Output spectral loss shows multiple loss peaks due to multiple SPRs between the core guided mode and SPP modes of different orders. This gives advantage for sensors based on metal-filled HC-PCF. This theoretical prediction has been experimented and the results show good sensitivity as attempted by different groups [32, 33].

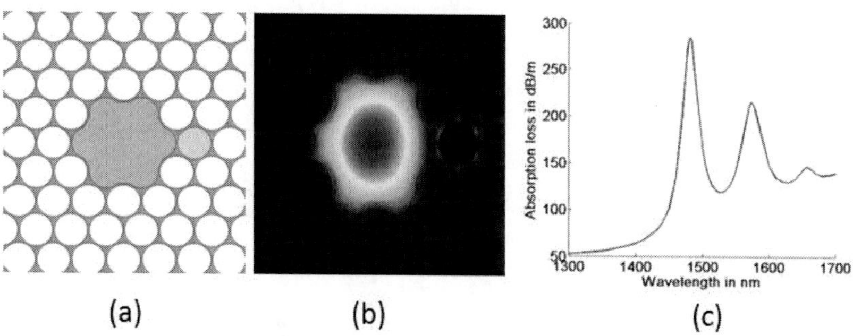

Figure 17. (a) Metal filled HC-PCF (hole on the RHS of hollow core), (b) Electric field distributions (FEM analyses) of metal-filled HC-PCF in presence of analyte in the core and (c) Absorption peaks of the SPP coupled core guided mode (Courtesy: Dr. Tushar Biswas, [31]).

Dirac Mode Transmission

A new type of HC-PCF structure is reported to achieve tunable Dirac mode transmission with unusual transmission characteristics. The concept of Dirac point occurring in electron energy band structure of graphene has been analogically utilized to achieve Dirac mode guidance in HC-PCF. Dirac point in the in-plane photonic band structure (PBS) of two-dimensional PC enabled localization of electromagnetic wave at Dirac frequency prevailing zero density of states [34]. At this point, a defect mode resonating at Dirac frequency is achieved by varying the defect radius.

It is demonstrated that a new type of light localization in a low index defect without PBG is observed due to formation of Dirac cones in PBS. The existence of Dirac point in out-of-plane PBS of triangular lattice PC along the longitudinal direction of specially designed HC-PCF is recorded by FDTD analysis. In this particular design GeO_2 doped high-index-composite-glass rings surrounding the air holes are considered to increase the effective dielectric contrast which is required to get Dirac cone in PBS. Results as shown in Figure 18 are encouraging; however, it needs to be experimentally demonstrated.

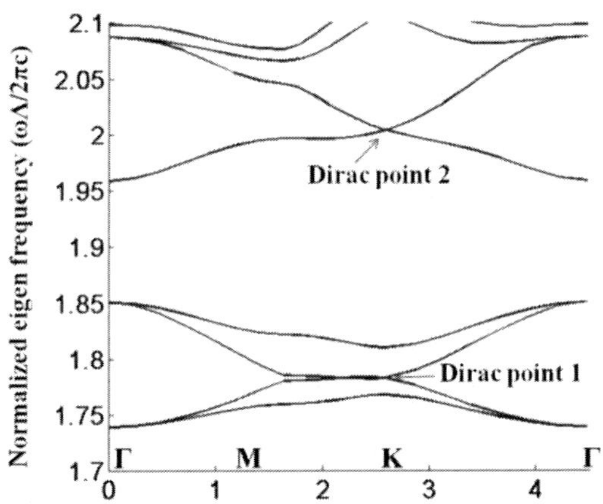

Figure 18. Out-of-plane PBS of HC-PCF with Dirac points calculated using the values of pitch (Λ) =2.21μm, n_{Si}=1.45, n_l=1.46. It is shown that band 2 and 3 degenerate at Dirac point 1 around **K** point. These bands overlap towards **MK** direction losing the conical dispersion relationship. Dirac point 2 occurring between bands 5 and 6 around **K** point maintains conical point. Linearity of these bands depends on the waveguide parameters.

Achieving linear dispersion at Dirac point would provide many niche applications. Some of the previous studies suggested the existence of Dirac point in out-of-plane PBS of dielectric/air triangular lattice PC [35, 36]. Dirac cones were observed using high index glass like Germania and SF6 glass. Fabrication of HC-PCFs using these soft glasses is very difficult due to smaller working temperature range. Besides, the viscosity in soft glasses changes very sharply with temperature. On the other hand, silica based HC-PCF is not suitable for Dirac frequency guidance as the Dirac cone degrades due to low RI contrast between host silica and air hole.

Dirac point in a triangular lattice of air holes surrounded by a Germania (GeO_2)-doped high-index ring in pure silica host (Figure 19) would be suitable for Dirac frequency guidance. The guidance wavelength can be tuned precisely by controlling the doping concentration and thickness of the doped glass rings. The unique frequency response of the Dirac mode can be compared with a resonant cavity mode where frequency selectivity is necessary.

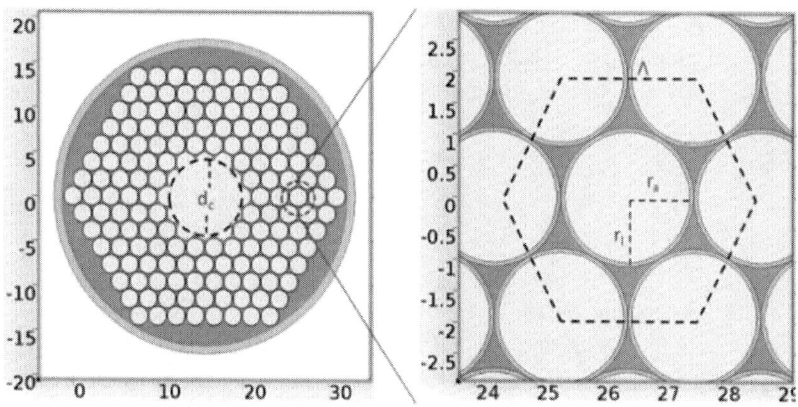

Figure 19. Schematic diagram of the cross section of the proposed HC-PCF for guiding Dirac frequency, thin light grey rings in the inset represent doped glass (n_l) (Courtesy: Dr. Tushar Biswas, Ref. [34]).

Conclusion

In this chapter the guidance mechanisms and the salient properties of conventional optical fiber as well as photonic crystal fibers (PCFs) are presented with special emphasis on the nonlinear processes and applications.

In order to understand the phenomenon of supercontinuum generation and other complex nonlinear phenomena occurring due to light-matter interaction in PCFs, the critical nature of dispersion needs to be understood clearly and that has been highlighted with necessary references.

As indicated in the chapter, the growth of the subject photonic crystals since 1987 has been phenomenal. It has opened up many areas of research including typical light guidance in photonic crystals and advanced topological metamaterials. Some of the theories and applications in this respect are highlighted. The development of PCFs, both solid and hollow core, in early 1990s has made tremendous impact in various applications such as supercontinuum light source and gas-filled lasers. Metal-filled hollow core PCF (HC-PCF) has shown great importance in sensor applications. All these niche applications and their development have been described with examples. Interesting futuristic new research area is also included such as Dirac mode guidance in HC-PCF for which special materials need to be explored. The entire content is supported by relevant references which may help the readers delve deeper into the subject.

Apart from nonlinear applications, HC-PCFs are increasingly foraying in the fields of quantum or coherent optics, metrology, spectroscopy and others. But these interesting application areas have not been surveyed here as it was outside the scope of this chapter.

Acknowledgements

The authors thank the Director, CSIR-Central Glass and Ceramic Research Institute, Kolkata, India and the Director, Indian Institute of Technology-Kharagpur (IIT-Kgp), Kharagpur, India for their unstinted support and encouragement.

References

[1] Yablonovitch, E. (1987). Inhibited spontaneous emission in solid-state physics and electronics. *Phys. Rev. Lett.*, 2059-2062.

[2] John, S. (1987). Strong localization of photons in certain disordered dielectric superlattices and electronics. *Phys. Rev. Lett.*, 2486-2489.

[3] Knight, J.C., Birks, T.A., Russell, P.St.J. and Atkin, D.M. (1996). All-silica single-mode optical fiber with photonic crystal cladding. *Opt. Lett.*, 1547-1549.

[4] Knight, J.C., Birks, T.A., Russell, P.St.J. and de Sandro, J.P. (1998). Properties of photonic crystal fiber and the effective index model. *J. Opt. Soc. Am. A*, 748-752.

[5] Kao, K.C. and Hockham, G.A. (1966). Dielectric-fibre surface waveguides for optical frequencies. *Proc. of IEEE*, 1151-1158.

[6] Payne, D.N. and Gambling, W.A. (1975). Zero material dispersion in optical fibres. *Electron. Lett.*, 176-178.

[7] Russell, P.St.J. (2003). Photonic crystal fibers. *Science*, 358-362.

[8] Snyder, A.W. and Love, J.D. 1983. *Optical Waveguide Theory*. New York: Chapman and Hall.

[9] Ghatak, A. and Thyagarajan, K. 1991. *Introduction to Fiber Optics*. New Delhi: Cambridge University Press.

[10] Bhadra, S.K. and Ghatak, A. 2013. *Guided wave optics and photonic devices*, Florida-USA: CRC Press.

[11] Keiser, G. 2008. *Optical Fiber Communications (4th Ed)*, New Delhi: Tata McGraw Hill Publishing Company Limited.

[12] Stolen, R.H. and Lin, C. 1978. Self-phase-modulation in silica optical fibers. *Phys. Rev. A*, 1448-1453.

[13] Islam, M.N., Mollenauer, L.F., Stolen, R.H., Simpson, J.R. and Shang, H.T. 1987. Cross-phase modulation in optical fibers. *Opt. Lett.*, 625-627.

[14] Stolen, R.H., Gordon, J.P., Tomlinson, W.J. and Haus, H.A. 1989. Raman response function of silica-core fibers. *J Opt. Soc. Am B*, 1159-1166.

[15] Inoue, K. and Toba, H. 1992. Wavelength conversion experiment using fiber four-wave mixing. *IEEE Photon. Tech. Lett.*,69-72.

[16] Agrawal, G.P. 2007. *Nonlinear Fiber Optics (4th Ed)*. Boston: Academic Press.

[17] Blow, K.J. and Wood, D. 1989. Theoretical description of transient stimulated Raman scattering in optical fibers. *IEEE J. Quantum. Electron.* 2665-2673.

[18] Gordon, J.P. 1986. Theory of the soliton self-frequency shift. *Opt. Letts.*, 662-664.

[19] Dudley, J.M., Genty, G. and Coen, S. 2006. Supercontinuum generation in photonic crystal fiber. *Rev. Mod. Phys.*, 1135-1184.

[20] Joannopoulos, J.D., Johnson, S.G., Winn, J.N. and Meade, R.D. 2011. *Photonic Crystals - Molding the Flow of Light (2nd Ed)*. New Jersey: Princeton University Press.

[21] Russell, P.St.J. 2006. Photonic-Crystal Fibers. *J. Lightwave Technol.*, 4729-4749.

[22] Benabid, F. 2006. Hollow core photonic band gap fiber: New light guidance for new science and technology. *Phil. Trans. R. Soc. A*, 3439-3462.

[23] Birks, T.A., Knight, J.C. and Russell, P.St.J. 1997. Endlessly single-mode photonic crystal fiber. *Opt. Lett.*, 961-963.

[24] Mogilevtsev, D., Birks, T.A. and Russell, P.St.J. 1998. Group-velocity dispersion in photonic crystal fibers. *Opt. Lett.*, 1662-1664.

[25] Knight, J.C., Arriaga, J., Birks, T.A., Ortigosa-Blanch, A., Wadsworth, W.J. and Russell, P.St.J. 2000. Anomalous dispersion in photonic crystal fiber. *IEEE Photon. Technol. Lett.*, 807 - 809.

[26] Reeves, W.H., Knight, J.C., Russell, P.St.J. and Roberts, P.J. 2002. Demonstration of ultra-flattened dispersion in photonic crystal fibers. *Opt. Exp.*, 609-613.

[27] Reeves, W.H., Skryabin, D.V., Biancalana, F., Knight, J.C., Russell, P.St.J., Omenetto, F.G., Efimov, A. and Taylor, A.J. 2003. Transformation and control of ultra-short pulses in dispersion-engineered photonic crystal fibres. *Nature*, 511-515.

[28] Ortigosa-Blanch, A., J.C. Knight, W.J. Wadsworth, J. Arriaga, B.J. Mangan, T.A. Birks, and P.St.J. Russell, 'Highly birefringent photonic crystal fibers', Opt. Letts. **25** 1325-1327 (2000).

[29] Kim, D. and Kang, J.U. 2004. Sagnac loop interferometer based on polarization maintaining photonic crystal fiber with reduced temperature sensitivity. *Opt. Exp.*, 4490-4495.

[30] Billotte, T., Chafer, M., Maurel, M., Amrani, F., Gerome, F., Debord, B. and Benabid, F. 2021. Contaminant-free end-capped and single-mode acetylene photonic microcell for sub-Doppler spectroscopy. *Opt. Lett.*, 456-459.

[31] Biswas, T., Chattopadhyay, R. and Bhadra, S.K. 2014. Plasmonic hollow-core photonic band gap fiber for efficient sensing of biofluids. *J. of Opt.*, 045001.

[32] Lu, Y., Yang, X., Wang, M. and Yao, J. 2015. Surface plasmon resonance sensor based on hollow-core PCFs filled with silver nanowires. *Electron. Lett.*, 1675-1677.

[33] Zhou, X., Cheng, T., Li, S., Suzuki, T. and Ohishi, Y. 2018. Practical sensing approach based on surface plasmon resonance in a photonic crystal fiber. *OSA Continuum*, 1332-1340.

[34] Biswas, T., Chattopadhyay, R. and Bhadra, S.K. 2016. Dirac-mode guidance in silica-based hollow-core photonic crystal fiber with high-index dielectric rings. *Phys. Status Solidi B*, 1898-1906.

[35] Xie, K., Jiang, H., Boardman, A.D., Liu, Y., Wu, Z., Xie, M., Jiang, P., Xu, Q., Yu, M. and Davis, L.E. 2014. Trapped photons at a Dirac point: a new horizon for photonic crystals. *Laser and Photonics Reviews*, 186-.

[36] Xie, K., Zhang, W., Boardman, A.D., Jiang, H., Hu, Z., Liu, Y., Xie, M., Mao, Q., Hu, L., Li, Q., Yang, T., Wen, F. and Wang, E. 2015. *Fiber guiding at the Dirac frequency beyond photonic bandgaps*. Light: Sc. & Appln., e304.

Chapter 4

Supercontinuum Generation in Photonic Crystal Fibers

Mário F. S. Ferreira*

I3N-Institute of Nanostructures, Nanomodelling and Nanofabrication, Department of Physics, University of Aveiro, Aveiro, Portugal

Abstract

The main physical mechanisms contributing to the supercontinuum generation process in photonic crystal fibers are discussed. Using a continuous wave or long duration pump pulses leads to SCG mainly due to the modulation instability phenomenon. Through this process the pump wave breaks up into fundamental solitons. A main contributor to the supercontinuum in this case is inelastic inter-soliton collisions. Pumping with femtosecond pulses in the anomalous dispersion region of the fiber, the soliton fission process gives origin to multiple fundamental solitons of different widths and peak powers. Among the host of soliton related effects, there are two which become then particularly important: the soliton self-frequency shift induced by the Raman scattering and the emission of dispersive radiation. Solitons dominate the long-wavelength edge of the supercontinuum, while the dispersive radiation determines the short wavelength expansion.

Keywords: supercontinuum generation, highly nonlinear fibers, photonic crystal fibers, optical solitons, dispersive waves

---

* Corresponding Author's Email: mfernando@ua.pt.

In: Photonic Crystal Fibers (PCFs)
Editor: Mário F. S. Ferreira
ISBN: 979-8-88697-618-2

1. Introduction

One of the most impressive nonlinear phenomena that can be observed in highly nonlinear fibers is the supercontinuum generation (SCG) [1]. It results generally from the synergy between several fundamental nonlinear processes, such as self-phase modulation, cross-phase modulation, stimulated Raman scattering, modulation instability, and four-wave mixing [2]. The spectral locations and powers of the pumps, as well as the nonlinear and dispersive characteristics of the medium determine the relative importance and the interaction between these nonlinear processes. Soliton related effects also play a significant role whenever light with sufficient power propagates in the anomalous dispersion regime [1]. A supercontinuum source can find applications in the area of bio-medical optics, where it allows the improvement of longitudinal resolution in optical coherence tomography by more than an order of magnitude, in optical frequency metrology, in all kinds of spectroscopy, in hyperspectral lidars, and as multiwavelength source in the telecommunications area.

In order to enhance the supercontinuum generation in optical fibers, it is important to increase the fiber nonlinear parameter by decreasing the effective mode area, together with the suitable adjustment of its dispersion characteristics. These developments have led to extensive research on photonic crystal fibers (PCFs) with optimized designs, namely using compound glasses [3-7]. The so-called jacketed air-suspended rod fiber [8] has the smallest possible effective mode area, but it is mainly a theoretical concept, which can only be approximately realized in practice. One of such approximations is provided by the triangular core fiber [4].

The supercontinuum generation process can be broadly divided in two categories, depending on the duration of the pump pulses. The first category corresponds to long (picosecond, nanosecond and continuous wave) pump pulses, whereas the second category is obtained with short (femtosecond) pump pulses. The key to long pulse and continuous wave (CW) SCG is the occurrence of modulation instability (MI), which is inherent to any anomalously dispersive, nonlinear medium [2]. Through this process the pump wave breaks up into fundamental solitons. The rate of breakup and the duration of these solitons depend on the pump conditions and fibre parameters such as dispersion and nonlinearity. The subsequent evolution of such solitons then leads to additional spectral broadening and supercontinuum formation through a variety of mechanisms [9-11].

Pumping in the normal dispersion regime using femtosecond pulses, where soliton formation is not allowed, leads only to spectral broadening through SPM and optical wave-breaking. In contrast, soliton dynamical effects play a prominent role when pumping in the anomalous dispersion regime using such ultrashort pulses. If the power of these pulses is high enough, they can evolve as higher-order solitons. However, because of perturbations such as higher-order dispersion or intrapulse Raman scattering, the dynamics of such pulses departs from the behavior expected of ideal high-order solitons and the pulses break up, giving origin to multiple fundamental solitons of different widths and peak powers [12]. Among the host of soliton related effects, there are two which become then particularly important: the soliton self-frequency shift induced by the Raman scattering and the emission of dispersive radiation [13, 14]. Solitons dominate the long-wavelength edge of the supercontinuum, while the dispersive radiation determines the short wavelength expansion [15].

In this Chapter we discuss the main physical mechanisms behind the generation of the supercontinuum in photonic crystal fibers, paying a special attention to the role played by soliton related phenomena in this process. An overview of the basic physics involved in supercontinuum generation and some main experimental results obtained by other authors are provided in Section 2. Modelling of the supercontinuum is discussed in Section 3, whereas some representative simulation results for different situations are presented in Section 4. A summary of the main conclusions is provided in Section 5.

2. Basic Physics of Supercontinuum Generation

Self-phase modulation (SPM) is one of the nonlinear effects which are responsible for the spectral broadening of short duration pulses. SPM-induced spectral broadening depends on the propagation distance, power density and pulse shape [2]. Its role is not significant at relatively low input powers and for relatively long pulses (within pico- and nanosecond range), so that the rate of power variation with time, dP/dt, is small. In fact, early experiments aiming the generation of supercontinuum in fibers revealed that it was due mainly to stimulated Raman scattering (SRS) and four-wave mixing (FWM) processes [16, 17]. Multimode propagation through the fiber provided the necessary phase matching conditions for the FWM process. Concerning the SRS process, its contribution for the supercontinuum generation is effective only on the

Stokes (long-wavelength) side of the spectrum. Hence, such contribution tends to make the overall spectrum asymmetric.

The four-wave mixing becomes progressively more important if the pumping is closer to the zero-dispersion wavelength (ZDW), since the parametric gain is higher than the Raman gain [2, 18]. Using pump pulses of sub-kilowatt peak power and a duration of some tens of ps, a spatially single-mode SC with more than 600 nm wide could be generated by the interplay of SRS and FWM [19, 20].

Actually, the global dispersive properties of the fiber become very important in the case of the FWM process. If the required phase-matching conditions are satisfied, FWM generates sidebands on the short- and long-wavelength sides of the pulse spectrum. This process can produce a wideband supercontinuum even in the normal dispersion regime of an optical fiber, where the positive nonlinear phase mismatch can be compensated for by the negative linear phase mismatch due to the fourth-order dispersion [2]. In a 1998 experiment, a SC spectrum over 140-nm was generated when 0.9-ps pulses were launched on a 1.7-km-long dispersion-flattened fiber having a normal GVD $b_2 = 0.1 \text{ps}^2/\text{km}$ at 1569 nm [21].

In conventional dispersion shifted fibers, the pulses are usually launched in the 1550 nm window and the resulting supercontinuum spectrum extends at most from 1300 to 1700 nm. However, photonic crystals fibers can be designed with a zero-dispersion wavelength (ZDW) of anywhere from 1550 down to 565 nm [2, 22]. Using such fibers, it becomes possible to generate a supercontinuum extending from the visible to the near infra-red region. Supercontinumm generation has been achieved in different PCFs pumping at 532 [23], 647 [24], 1064 [25], and 1550 nm [26]. Actually, besides the possibility of adjusting the ZDW, the remarkable features of PCFs provides a plethora of nonlinear mechanisms of spectral broadening that may differ depending on waveguide dispersion profile, power, wavelength and duration of pumping pulses.

Pumping in the anomalous dispersion regime using picoseconds pump pulses, the spontaneous modulation instability (MI) becomes the dominant process in the initial propagation phase. In the frequency domain, the modulation instability causes the spread of energy from the central wave frequency to side-bands with a frequency separation given by [2, 18]:

$$\Omega_s = \sqrt{\frac{2\gamma P_p}{|\beta_2|}} \tag{1}$$

where b_2 is the group-velocity dispersion, g is the fiber nonlinear parameter, and P_p is the pump peak power. The MI period can be obtained from the above frequency separation as

$$T_s = \frac{2p}{W_s} = \sqrt{\frac{2p^2 |b_2|}{g P_p}} \tag{2}$$

The modulated picoseconds pulse can then break up into a train of solitons with durations approximately given by T_s. The subsequent evolution of these solitons then leads to additional spectral broadening and supercontinuum formation through a variety of mechanisms. If such solitons are not sufficiently short, they can undergo soliton collisions, which determine the transfer of energy among them. If the solitons resulting from the initial pulse break up are sufficiently short, they can be affected by some higher-order dispersive and nonlinear effects. For example, in fibers with positive dispersion slopes, these soliton pulses shed energy to phase-matched nonsoliton radiation (NSR) located on the short-wavelength side of the pump [14]. Simultaneously, they are also affected by the intrapulse Raman scattering (IRS) phenomenon [9, 27]. This leads to a continuous downshift of the soliton carrier frequency, an effect known as the soliton self-frequency shift (SSFS) [28]. The origin of SSFS can be understood by noting that for ultrashort solitons the pulse spectrum becomes so broad that the high-frequency components of the pulse can transfer energy through Raman amplification to the low-frequency components of the same pulse. Such an energy transfer appears as a red shift of the soliton spectrum, with shift increasing with distance. Optical rogue waves can be observed in association with red-shifted solitons with uncommon probability distributions arising from picosecond SC generation [29-31].

The general features observed in the case of long duration pulses can also be extended to the continuous wave pumping in the anomalous dispersion regime. In this case, the supercontinuum generation also starts with the modulation instability, which leads to the splitting of the continuous wave into multiple fundamental solitons. The modulation instability process evolves from background noise and is often stimulated by the intensity fluctuations which are typical of a continuous wave pump [32-34].

In some circumstances, the fundamental solitons arising from MI can be short enough in order to enable the occurrence of intrapulse Raman scattering, leading to the soliton self-frequency shift. As the modulation instability evolves from noise, there will be some jitter on the period and hence also on the duration of the generated solitons. In this case, a range of soliton self-frequency shifts occurs, creating a smooth red-shifted continuum evolution.

Soliton collisions constitute also an important mechanism for the spectral expansion in CW pumped supercontinua [10, 11]. In this case, the number of solitons can be particularly large and as red shifted solitons travel more slowly than those at higher frequencies, they can eventually collide while propagating along the fiber. During such collisions, some energy will be transferred to the most red-shifted soliton, if it is within the spectral range of Raman scattering. The energy increase leads to a corresponding compression, which results in a further red-shift for that soliton. The increased red-shift also slows down the soliton, leading it to pass by more solitons and hence experience more collision events. Soliton collisions can considerably increase the extent of a supercontinuum compared to Raman self-scattering alone.

The expansion of the supercontinuum to longer wavelengths can be limited by either the dispersion magnitude increasing or nonlinearity decreasing. Depending on the dispersion profile, the existence of a second ZDW can lead to the generation of dispersive waves and the spectral recoil of the solitons [35, 36]. Concerning the extension to shorter wavelengths, it is determined by the excitation of phase-matched dispersive waves and soliton trapping mechanisms. Experimentally, this phenomenon has been combined with dispersion-engineered PCFs to further extend SCs to the UV in nanosecond and picosecond pumping schemes [37].

In a 2008 experiment a SC ranging from 670 nm to 1350 nm with 9.55W output power was achieved in the CW pumping regime, using a 200-m long ZDW decreasing MOF pumped by a 20 W ytterbium fiber laser at 1.06 μm [38]. The fiber was composed by a 100 m long section with a constant dispersion followed by a 100m long section with linearly decreasing ZDW. The initial ZDW was located at 1053 nm, just below the pump wavelength and decreased to 950 nm in a quasi-linear way.

Pumping a PCF with femtosecond pulses in the normal dispersion regime, where soliton formation is not allowed, leads only to spectral broadening through SPM [39, 40]. As the intensity of the input pulse is increased, the spectrum is increasingly broadened. If the pump wavelength is located near the zero-dispersion wavelength, the spectrum can extend into the anomalous dispersion regime. In such case, the soliton dynamics can have a significant contribution to the spectral broadening [41].

Actually, soliton dynamical effects play a prominent role when pumping in the anomalous dispersion regime using femtosecond pulses. If the power of the pump pulses is high enough, they can evolve as higher-order solitons. These pulses begin experiencing a spectral broadening and temporal compression which are typical of higher-order solitons [2, 18]. However, because of perturbations such as higher-order dispersion or intrapulse Raman scattering, the dynamics of such pulses departs from the behavior expected of ideal high-order solitons and the pulses break up. This process is known as soliton fission and determines the separation of each higher-order soliton pulse into several fundamental solitons [42].

The soliton fission process is affected if the input pulse wavelength is near a ZDW. In such case the ejected solitons shed some of their energy to dispersive waves, generated on short-wavelength side of ZDW in fibers with positive dispersion slopes [43, 44]. Since the dispersive waves have a lower group-velocity, they will lag behind the solitons. However, the group velocity of the red-shifting solitons decreases during the propagation and eventually they meet with the dispersive waves. The temporal overlap between a soliton and a dispersive wave allows them to interact via the XPM effect, which determines a blue-shift and a decrease in group-velocity of the dispersive wave. In this way, as long as this cycle continues, the blue-shifted radiation cannot escape from the soliton, which is usually referred as a "trapping effect" [45, 46]. The XPM effect determines the spectral broadening of the dispersive waves and merging of different peaks, helping in the formation of a broadband supercontinuum [47-51].

PCFs based on silica are particularly attractive candidates for high-power SC generation in the visible and near infrared (NIR) regions, which require low-loss materials in the wavelength range of interest, as well as appropriate nonlinear coefficient and controllable dispersion. SC sources in these regions have attracted considerable attention due to potential applications in optical frequency metrology, spectroscopy, hyperspectral lidar, medicine and biomedicine, and optical communications area [52-61]. In a 2016 experiment, a picosecond ytterbium-doped fiber laser has been used to pump a piece of 5 m long PCF to obtain 67.9 W SC with spectrum coverage of 500-1700 nm [62]. In 2018, an 80 W SC with spectra covering 350-2400 nm from multicore PCFs was demonstrated [63]. In the same year, a PCF SC source with 215 W of average power and spectra spanning from 480 nm to over 2000 nm was reported [64]. A linearly polarized SC with 93 W average output power and spectrum ranging from 520 nm to 2300 nm was demonstrated in a 2019 experiment using a piece of 2.6 m long polarization maintaining PCF and a picosecond pump source [65].

Actually, most of the common silica fiber SC sources are limited on the short-wavelength edge, normally around 450 nm [66]. However, a blue-enhanced supercontinuum can be achieved efficiently using tapered or specially designed PCFs. In a 2011 experiment, blue-extended SC spectra down to 372 nm in simply designed high air-filling fraction (> 0.9) PCFs was reported [67]. Later, a flat ultraviolet-extended SC down to 352 nm pumped at 1064 nm in cascading PCF tapers has been reported [68]. More recently, an ultraviolet-enhanced SC down to 350 nm was obtained using a specifically designed seven-core PCF pumped by a picosecond Yb-doped master oscillator power amplifier (MOPA) [69].

Besides the above mentioned limitations on the short wavelength range, the material loss in silica increases drastically at wavelengths beyond ~2μm, effectively preventing the spectral evolution of SCG into the mid-infrared (MIR) region, where many molecular bonds have fundamental vibration frequencies, [70]. MIR SC light sources show remarkable potential in various applications, such as tomography [71], food quality monitoring [72], disease diagnosis [73], and national defense [74]. Different types of glasses have been used to make fibers able to generate MIR SC, namely tellurite [75, 76], fluoride [77, 78], and chalcogenide (ChG) glasses [79, 80]. Compared with other glasses, ChG glass exhibits a wider MIR transmission window up to 20 μm and a higher optical nonlinearity up to three orders of magnitude higher than that of silica glasses [81], making them promising candidates for MIR SC generation.

In recent years, many ultrabroadband SCs in step-index ChG fibers have been generated by pumping in the anomalous dispersion regime [82]. In a 2014 experiment, an ultrabroad MIR SC spanning from 1.4 to 13.3 μm using an 85-mm-long step-index fiber made of chalcogenide fiber pumped at 6.3 μm was demonstrated [83]. A broadband SC covering 2-12 μm spectral range in a 11-cm-long Ge-Sb-Se fiber pumped at 4.485 μm with 330 fs pulse was reported in 2016 [84]. In another 2016 experiment, a MIR SCG spanning ~2.0 to 15.1 μm in a 3 cm-long chalcogenide step-index fiber pumped at 9.8 μm with 170 fs pulses has been demonstrated [85].

In all cases mentioned above, SC is generated by pumping in the anomalous dispersion regime and it is mainly due to soliton dynamics, such as soliton fission and self-frequency shift. However, this broadening process generates a spectrum with a complex temporal profile, which is sensitive to noise amplification, leading to different spectral components [86]. These features cause coherent degradation of SCs and reduce resolution or precision in many applications. A common approach to avoid these limitations is to pump fibers with an all-normal dispersion profile by using ultrashort pulse lasers [87-89]. In these cases, the fiber exhibits flattened normal dispersion over the entire spectral region of interest, and the SC spectral broadening is mainly as a result of self-phase modulation and optical wave breaking [90], which can eliminate noise-sensitive soliton dynamics [91-93].

3. Modelling of the Supercontinuum

Modelling the supercontinuum generation can be realized considering a generalized nonlinear Schrödinger equation (NLSE) that includes higher-order dispersion effects, as well as intrapulse Raman scattering. Such equation can be written as [1, 2, 18].

$$\frac{\partial U}{\partial z} - i\sum_{k\geq 2}\frac{i^k \beta_k}{k!}\frac{\partial^k U}{\partial t^k} + \frac{\alpha(\omega)}{2}U = i\gamma\left(1 + \frac{i}{\omega_0}\frac{\partial}{\partial t}\right)\left(U(z,t)\int_{-\infty}^{+\infty} R(t')\left|U(z,t-t')\right|^2 dt'\right) \tag{3}$$

where $U(z,t)$ is the electric field envelope, ω_0 is the center frequency, β_k are the dispersion coefficients at the center frequency, $\alpha(\omega)$ is the frequency-dependent fiber loss, and γ is the nonlinear parameter.

The nonlinear response function $R(t)$ in Eq. (3) can be written as

$$R(t) = (1 - f_R)\delta(t) + f_R h(t), \tag{4}$$

where the δ-function represents the instantaneous electron response (responsible for the Kerr effect), $h(t)$ represents the delayed ionic response (responsible for the Raman scattering) and f_R is the fractional contribution of the delayed Raman response to the nonlinear polarization, in which a value $f_R = 0.18$ is often assumed [94]. It is common to approximate $h(t)$ in the form [18, 95]:

$$h(t) = \frac{t_1^2 + t_2^2}{t_1 t_2^2} \exp(-t/t_2) \sin(t/t_1) \tag{5}$$

where $t_1 = 12.2$ fs and $t_2 = 32$ fs. More accurate forms of the response function $h(t)$ have also recently been investigated [96].

Eq. (3) can be used to describe the propagation of femtosecond pulses in optical fibers, in both the normal and anomalous dispersion regimes. When such pulses have enough power, their spectra undergo extreme broadening. In the anomalous dispersion regime, this process is mainly influenced by the phenomenon of soliton fission, which occurs whenever a higher-order soliton is affected by third- or higher-order dispersion. The soliton order N is given by [2, 18]:

$$N = \sqrt{\frac{L_D}{L_{NL}}} \tag{6}$$

where

$$L_D = \frac{t_0^2}{|b_2|} \tag{7}$$

is the dispersion distance and

$$L_{NL} = \frac{1}{g P_0} \tag{8}$$

is the nonlinear length.

In the presence of higher-order dispersion, an Nth-order soliton gives origin to N fundamental solitons whose widths and peak powers are given by [97]:

$$t_k = \frac{t_0}{2N + 1 - 2k} \tag{9}$$

$$P_k = \frac{(2N+1-2k)^2}{N^2} P_0 \tag{10}$$

where $k = 1$ to $\tilde{N}$, where $\tilde{N}$ is the integer closest to N when N is not an integer. Soliton fission occurs generally after a propagation distance

$$L_{fiss} \sim L_D / N , \tag{11}$$

at which the injected higher-order soliton attains its maximum bandwidth. The fission distance L_{fiss} is a particularly significant parameter in the context higher-order soliton effect compression [98, 99]

The higher-order soliton fission process is also affected by the IRS phenomenon, which leads to a red-shift of the generated fundamental solitons. The rate of frequency shift per propagation length is given by [2]

$$\frac{df}{dz} = -\frac{4t_R|\beta_2|}{15\pi t_0^4} = -\frac{4t_R(\gamma P_0)^2}{15\pi|\beta_2|} \tag{12}$$

where $t_R \equiv f_R \int_{-\infty}^{+\infty} t h(t)dt \approx 5\,\text{fs}$ is the Raman parameter and P_0 is the soliton peak power. Since the SSFS effect is proportional to $(\gamma P_0)^2$, it will be enhanced if short pulses with high peak pulses are propagated in highly nonlinear fibers.

The higher-order soliton evolution and soliton fission as described above become progressively less important when pumping in the anomalous dispersion regime with long duration and high peak power pulses. This is because the fission distance L_{fiss} also increases with the pump pulse duration. Under these conditions, the modulation instability (MI) causes the temporal breakup of the pump pulse into a large number of fundamental solitons. The subsequent evolution of these solitons then leads to additional spectral broadening and supercontinuum formation through a variety of mechanisms, namely soliton dynamical effects such as dispersive wave generation and the Raman-induced frequency shift.

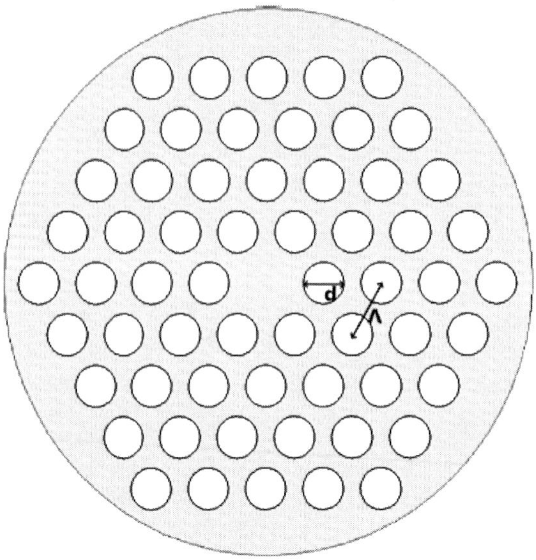

Figure 1. Schematic representation of a conventional photonic crystal fiber.

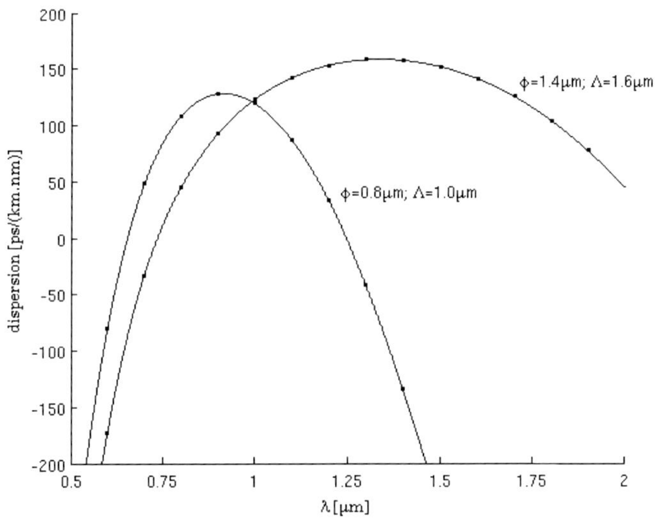

Figure 2. Dispersion curves for two different photonic crystal fibers with a hexagonal pattern of holes: fiber A, with holes of diameter $d = 1.4$ μm spaced by $L = 1.6$ μm, and fiber B with holes of diameter $d = 0.8$ μm, spaced by $L = 1$ μm.

4. Simulation Results

In this Section we present and discuss several simulation results for the supercontinuum generation in photonic crystal fibers. Pumping in the anomalous dispersion regime using femtosecond pulses is assumed, such that the soliton dynamical effects play a prominent role.

Figure 1 illustrates a conventional solid-core photonic crystal fiber (PCF), which has a set of holes arranged in a hexagonal lattice. This fiber has two main structural parameters: the air-hole diameter d and the hole-to-hole spacing L .

Figure 2 shows the dispersion curves for two different conventional PCFs: fiber A, with holes of diameter $d = 1.4$ μm spaced by L $= 1.6$ μm, and fiber B with holes of diameter $d = 0.8$ μm spaced by L $= 1$ μm. Fiber A has is a single zero dispersion wavelength at 735 nm, whereas fiber B has two ZDWs: one at 654 nm and the other at 1248 nm

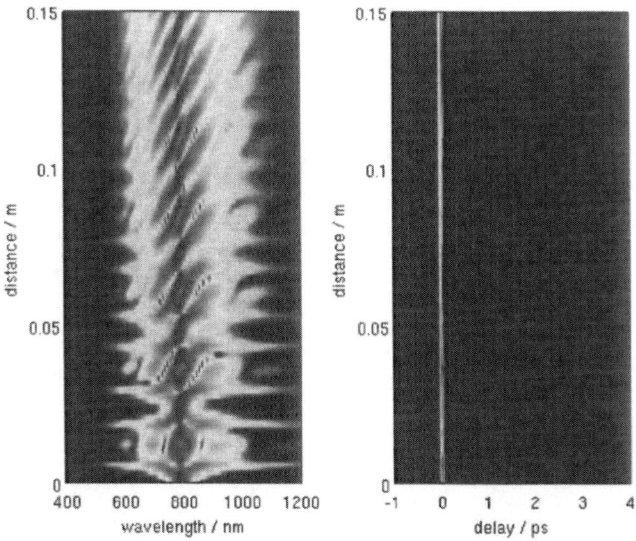

Figure 3. (a) Spectral and (b) temporal evolution of an optical pulse along a photonic crystal fiber obtained from Eq. (3) in the presence of 2nd order dispersion, self-phase modulation (SPM) and self-steepening (SST) effects. The input sech pulse at 790 nm has a peak power of 5 kW and 25 fs FWHM.

Figure 3 shows the temporal and spectral evolution of an optical pulse along the photonic crystal fiber A. This result was obtained by numerically solving the Eq. (3), taking into account only the 2nd order dispersion, self-phase modulation (SPM) and self-steepening (SST) effects. The pump was assumed in the anomalous dispersion regime at 790 nm, where the nonlinear parameter is $g = 117\,\mathrm{W}^{-1}\mathrm{km}^{-1}$. An input pulse $U(0,t) = \sqrt{P_0}\,\mathrm{sech}(t/T_0)$, is assumed, where $P_0 = 5$ kW, and the width $T_0 = 14.2$ fs, which corresponds to an intensity full width at half maximum FWHM of 25 fs. For the assumed pulse and fiber parameters, we have N = 3.01 and a fission distance $L_{fiss} \sim 0.51$ cm.

We note in Figure 3 a spectral broadening of the input pulse, which does not exhibit a periodic evolution. The impact of intrapulse Raman scattering (IRS) on pulse propagation can be seen in Figure 4, where we may observe the soliton fission and the three individual pulses resulting from this process. These solitons are red-shifted due to the Raman effect, being called Raman solitons. The change of the soliton's frequency leads to a bending of the soliton trajectory in the time domain.

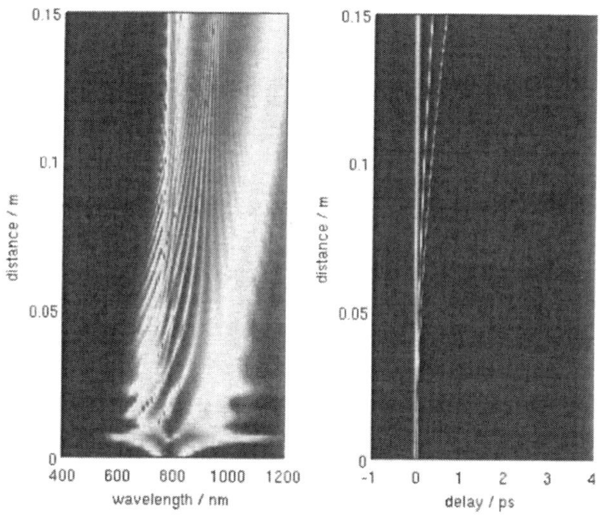

Figure 4. (a) Spectral and (b) temporal evolution of an optical pulse along a photonic crystal fiber obtained from Eq. (3) in the presence of 2nd order dispersion, self-phase modulation (SPM), self-steepening (SST), and intrapulse Raman scattering (IRS) effects. The fiber and pump conditions are the same as for Figure 3.

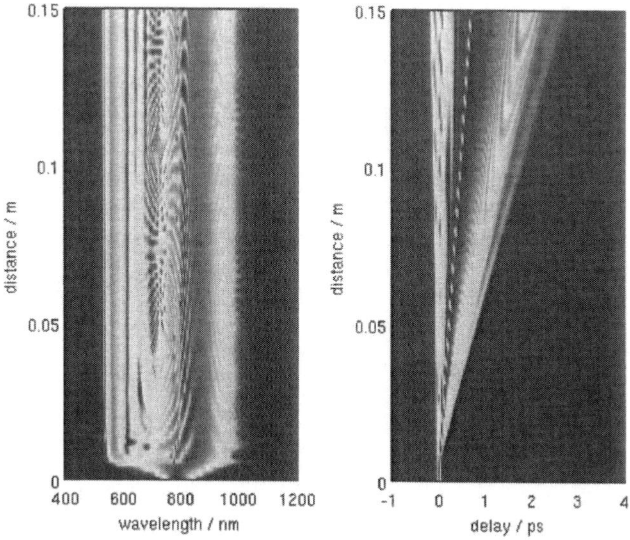

Figure 5. (a) Spectral and (b) temporal evolution of an optical pulse along a photonic crystal fiber obtained from Eq. (3) taking into account the higher-order dispersion, SPM, and SST effects, but neglecting IRS. The fiber and pump conditions are the same as for Figure 3.

Figure 5 shows the temporal and spectral evolution of the optical pulse along taking into account third- and higher-order dispersion, SPM and SST effects, but neglecting the intrapulse Raman scattering. The individual solitons resulting from the fission process are not apparent in Figure 5 because they do not separate much from each other when the Raman contribution is neglected. However, a clear signature of soliton fission is the appearance of a new spectral peak in Figure 5a at $z = L_{fiss}$. Regarding the time domain, Figure 5b shows that a dispersive wave spreads with propagation and lags behind the main soliton.

The new spectral peak observed in Figure 5a corresponds to the so-called *nonsoliton radiation* (NSR) [100] or *Cherenkov radiation* [14] which is emitted by the solitons resulting from the fission process in the presence of higher-order dispersion. The NSR is emitted at a frequency such that its phase velocity matches that of the soliton. Neglecting the fourth- and higher-order dispersions, the frequency shift $\Omega_d = \omega_d - \omega_s$ between the frequency of dispersive waves (ω_d) and that of the soliton (ω_s) is given approximately by [14]:

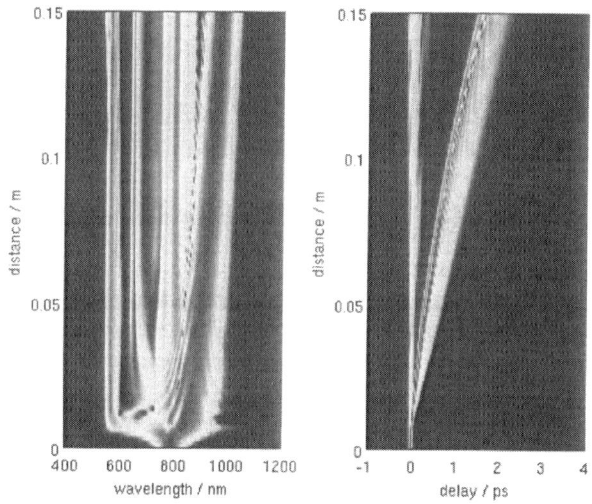

Figure 6. (a) Spectral and (b) temporal evolution of an optical pulse along a photonic crystal fiber obtained from Eq. (1) taking into account both the higher dispersion and the Raman effect. The fiber and pump conditions are the same as for Figure 3.

$$W_d \gg -\frac{3b_2}{b_3} + \frac{gP_0b_3}{3b_2^2} \tag{16}$$

where b_j (j = 2, 3) are the dispersion parameters, calculated at the soliton central frequency w_s. For solitons propagating in the anomalous-GVD regime we have $b_2 < 0$. In these circumstances, Eq. (15) shows that $W_d > 0$ when $b_3 > 0$, in which case the NSR is emitted at wavelengths shorter than that of the soliton, as observed in Figure 7a. On the contrary, if $b_3 < 0$ the NSR is emitted at a longer wavelength than that of the soliton.

Figure 6 shows the spectral and temporal evolution of the optical pulse along a photonic crystal fiber obtained from Eq. (1) taking into account both the full dispersion curve and the Raman effect. Figure 7 shows the initial and final pulse profiles, both in frequency and time domains.

The most noticeable difference of Figure 6 in comparison with Figure 5 is the red-shift of the solitons created by the fission process. As a consequence of the SSFS effect induced by intrapulse Raman scattering, such solitons separate from each other. According with Eq. (7), since the SSFS is the largest for the shortest soliton, its spectrum shifts the most toward the red side in

Figure 6a. The change of the soliton´s frequency determines a reduction in the soliton´s speed because of dispersion. This deceleration appears as a bending of the soliton trajectory in the time domain, as observed in Figure 6b.

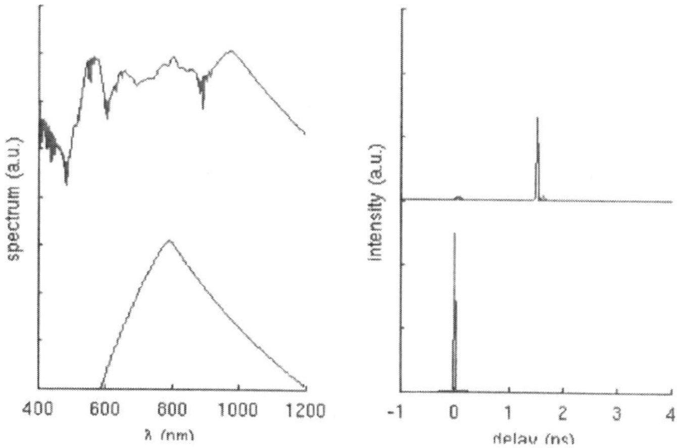

Figure 7. Initial and final pulse profiles in (a) frequency and (b) time domains for the case of Figure 6.

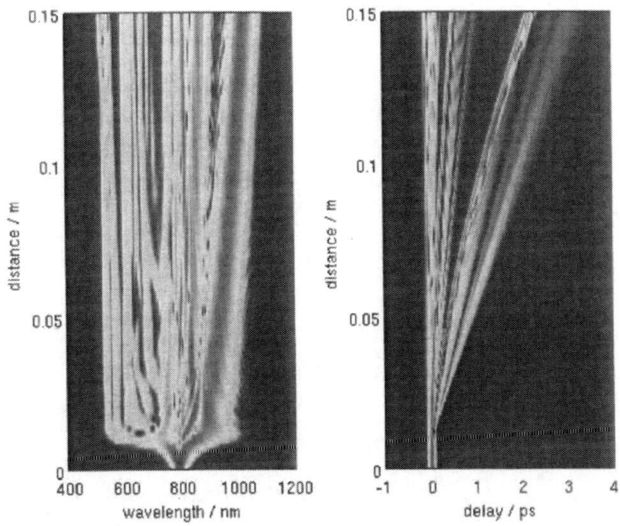

Figure 8. (a) Spectral and (b) temporal evolution of an optical pulse along a photonic crystal fiber obtained from Eq. (1). The input sech pulse at 790 nm has a peak power of 5 kW and 50 fs FWHM.

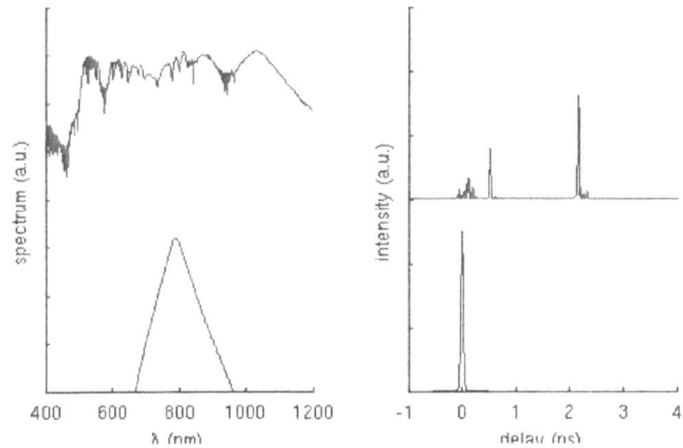

Figure 9. Initial and final pulse profiles in (a) frequency and (b) time domains for the case of Figure 8.

Keeping the input pulse peak power at 5 kW but increasing its width to 50 fs provides a soliton order $N = 6.02$. Figure 8 shows that, in this case, we achieve a wider and more uniform supercontinumm in comparison with that of Figure 6. Figure 9 shows the initial and final pulse profiles, both in frequency and time domains. Since the power of the first ejected soliton is greater than in the case of Figure 6, it experiences a larger SSFS, according with Eq. (7). At the same time, the corresponding nonsolitonic radiation is generated at a lower wavelength in the normal region.

As seen in Figure 2, fiber B has two ZDWs, such that the GVD is anomalous in the spectral window between them. The third-order dispersion parameter b_3 is positive near the first ZDW and negative near the second one. Near the first ZDW the NSR is always blue-shifted, as observed in Figures 6 and 8, so that the recoil and SSFS act in the same spectral direction. However, the opposite behaviour must occur near the second ZDW. This situation is illustrated in Figure 10, which shows the supercontinuum generated in fiber B, when it pumped at 860 nm using 100 fs pulses with a peak power of 1 kW. In this case, the GVD at the pump wavelength is $b_2 = -0.0487$ ps^2/m and the nonlinear parameter is $g = 196.2\ \text{W}^{-1}\text{km}^{-1}$, which provide a soliton order N = 3.6. As the soliton loses energy to the red-shifted NSR, which is emitted in the normal-GVD regime, it becomes slightly wider in the time domain, as a result of energy conservation. On the other hand, since the detuning between the radiation and soliton gets smaller with propagation (Figure 10), the radiation is exponentially amplified. This amplification leads to the strong spectral recoil on the soliton into the anomalous-GVD regime, providing the suppression of SSFS [36, 101, 102]. These effects are clearly visible in Figure 10, regarding the evolution of the first ejected soliton. We observe that the frequency shift of this soliton is suppressed after Z ~ 0.4 m.

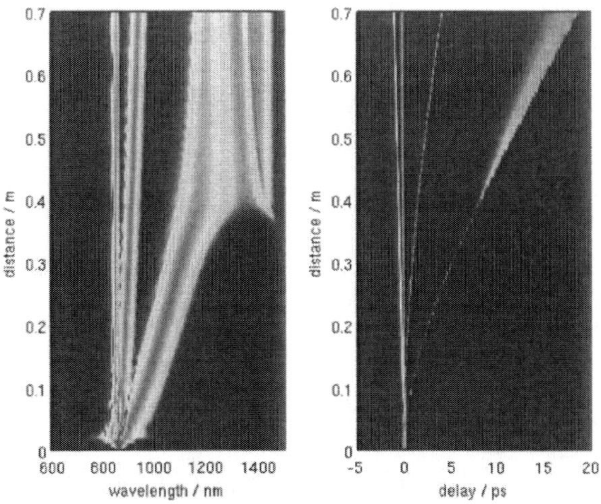

Figure 10. (a) Spectral and (b) temporal evolution of the supercontinuum generated in fiber B of Figure 6, pumped at 860 nm using 100 fs pulses with a peak power of 1 kW.

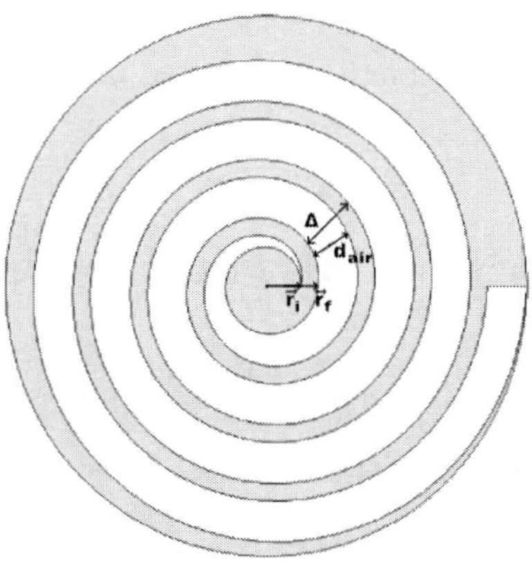

Figure 11. Schematic representation of the LS-MOF, including its three main parameters: *ri*, *dair* and Δ. The gray region corresponds to glass, and the white region to air.

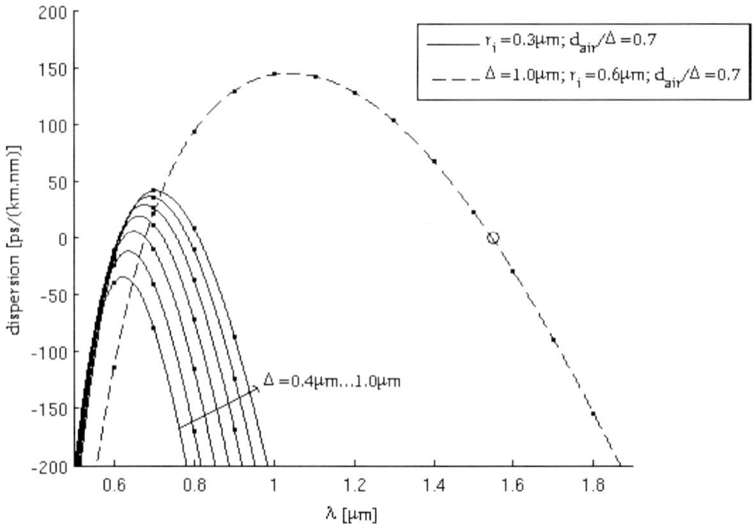

Figure 12. Dispersion curves of the LS-MOF corresponding to the indicated parameters.

A rather interesting and nonconventional PCF, which can provide a particularly high nonlinear parameter, is the so-called layered spiral microstructured optical fiber (LS-MOF) [103]. In the design of the LS-MOF there are two archimedean spirals and one equiangular spiral, both spinning around the center, as illustrated in Figure 11. There are three relevant parameters in this structure: *ri* which is the radius of the circle enclosed in the LS-MOF, *dair*/Δ which is approximately equal to the fraction of air in the cladding, and Δ which controls the localization of the regions of glass and air. In the limit $d_{air}/\Delta\rightarrow 1$ the LS-MOF tends to the jacketed air-suspended rod fiber [8].

With a careful adjustment of the LS-MOF parameters, a zero dispersion can be found at 1.546μm~1.55μm, for *ri* = 0.6μm, *dair*/Δ = 0.7 and Δ = 1.0 μm as discussed in ref. [103]. A zero dispersion near 0.650μm for *ri* = 0.3μm, *dair*/Δ = 0.7 and Δ = 0.6μm. The dispersion corresponding to these situations is shown in Figure 12, where the circle corresponds to the zero dispersion point at 1.55μm.

Figure 13 illustrates the supercontinuum generation in the case of a LS-MOF with the parameters *ri* = 0.6μm, *dair*/Δ = 0.7 and Δ = 1.0μm. The corresponding curve for the dispersion is presented in Figure 12, which exhibits two zero dispersion wavelengths: one at 679 nm, and the other at 1546 nm. Pumping was assumed at 780 nm (anomalous dispersion region), using 50 fs pulses with a peak power of 10 kW. The corresponding nonlinear parameter has a value of 187.1 $W^{-1}km^{-1}$. We can notice that the spectrum is significantly broadened both to lower and to higher values relative to the pumping wavelength, as a result of the soliton dynamics and dispersive wave generation. The initial pulse corresponds to a soliton of order $N = 7.1$, which desintegrates under the soliton fission phenomenon., The supercontinuum illustrated in Figure 13 extends approximately from 700nm to1200nm.

A record value of the nonlinear parameter of 50.7 $W^{-1}m^{-1}$ at 1.550μm was achieved by optimizing the design of a LS-MOF constituted by arsenic trisulfide. Using such optimized design, a broad supercontinuum, extending from 500 nm to 3900 nm, was observed after a pulse propagation of only 0.3mm [104]

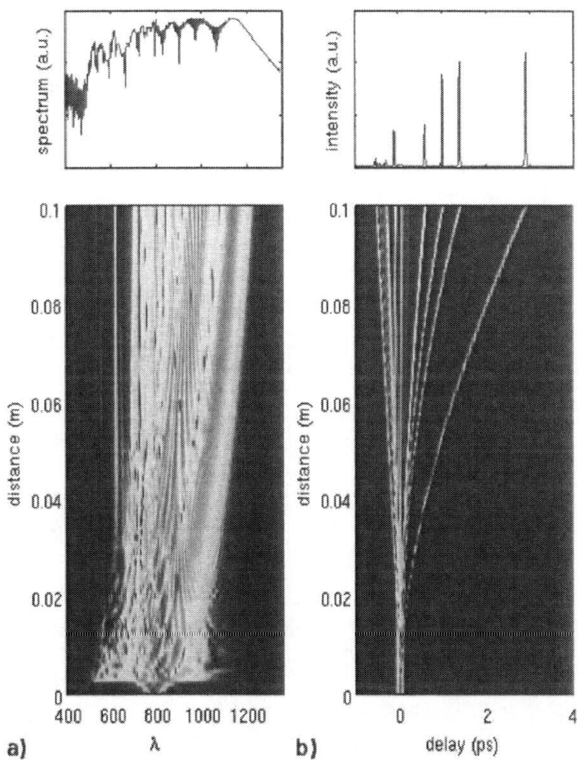

Figure 13. (a) Spectral and (b) temporal evolution of the supercontinuum generated in a LS-MOF using 50 fs pumping pulses at 780 nm.

5. Conclusion

In this chapter we discussed the main physical mechanisms contributing to the supercontinuum generation process in photonic crystal fibers. This process can be broadly divided in two categories, depending on the duration of the pump pulses. Using long (picosecond, nanosecond and continuous wave) pump pulses leads to SCG mainly due to the modulation instability (MI) phenomenon, which is inherent to any anomalously dispersive. Through this process the pump wave breaks up into fundamental solitons. The duration of such solitons can be short enough to have the spectral width necessary for Raman self-scattering, leading to the soliton self-frequency shift. Another main contributor to the supercontinuum in this case is inelastic inter-soliton collisions. Pumping with femtosecond pulses in the anomalous dispersion region of the fiber, the soliton fission process gives origin to multiple fundamental solitons of different widths and peak powers. Among the host of soliton related effects, there are two which become then particularly important: the SSFS induced by the Raman scattering and the emission of dispersive radiation. Solitons dominate the long-wavelength edge of the supercontinuum, while the dispersive radiation determines the short wavelength expansion.

References

[1] Dudley, J., and Taylor, J. (Eds) (2010). *Supercontinuum Generation in Optical Fibers*. Cambridge University Press, Cambridge, UK.
[2] Ferreira, M. F. (2011). *Nonlinear Effects in Optical Fibers*, John Wiley & Sons.
[3] Russel, P. (2006). *J. Lightwave Technol.*, 24, 4729.
[4] Leong, J., Petropoulos, P., Price, J., Ebendorff-Heidepriem, H., Asimakis, S., Moore, R., Frampton, K., Finazzi, V., Feng, X., Monro, T., and Richardson, D. (2006). *J. Lightwave Technol.*, 24, 183.
[5] Ebendorff-Heidepriem, H., Petropoulos, P., Asimadis, S., Finazzi, V., Moore, R., Frampton, K., Koizumi, F., Richardson, D., and Monro, T. (2002). *Opt. Express*, 12, 5082
[6] Ravi, V., Kumar, K., George, A., Reeves, W., Knight, J., Russell, P., Omenetto, F., and Taylor, A. (2002). *Opt. Express*, 10, 1520.
[7] Ravi, V., Kumar, K., George, A., Knight, J., and Russel, P. (2003). *Opt. Express*, 11, 2641.
[8] Finazzi, V., Monro, T., and Richardsom, D. (2003). *J. Opt. Soc. Am. B.*, 20, 1427.
[9] Mitschke, F., and Mollenauer, L. (1986). *Opt. Lett.*, 11, 659.

[10] Korneev, N., Kuzin, E. A., Ibarra-Escamilla, B., and Flores-Rosas, M. B. J. A. (2008). *Opt. Express*, 16, 2636.
[11] Frosz, M. H., Bang, O., and Bjarklev, A. (2006). *Opt. Express*, 14, 9391.
[12] Kodama, Y., and Hasegawa, A. (1987) *EEE J. Quant. Elect.*, QE23, 510.
[13] Wai, P. K. A., Menyuk, C. R., Lee, Y. C., and Chen, H. H. (1986). *Opt. Lett.*, 11, 464
[14] Akhmediev, N., and Karlsson, M. (1995). *Phys. Rev. A*, 51, 2602.
[15] Dudley, J. M., Genty G., and Coen, S. (2006). *Rev. Mod. Phys.*, 78, 1135.
[16] Lin, C., and Stolen, R. H. (1976). *Appl. Phys. Lett.*, 28, 21.
[17] Baldeck, P., and Alfano, R. (1987) *J. Lightwave Technol.*, 5, 1712.
[18] Agrawal, G. P. (2007). *Nonlinear Fiber Optics*, 4th Ed. San Diego: Academic Press.
[19] Coen, S., Chau, A., Leonhardt, R., Harvey, J., JKnight, J., Wadsworth, W., and Russel, P. (2002). *J. Opt. Soc. Am. B*, 19, 753.
[20] Dudley, J., Provino, L., Grossard, N., Maillotte, H., Windeler, R., Eggleton, B., and Coen, S. (2002). *J. Opt. Soc. Am. B*, 19, 765.
[21] Takushima, Y., Futami, F., and Kikuchi, K. (1998). *IEEE Photon. Technol. Lett.*, 10, 1560.
[22] Reeves, W., Skryabin, D., Biancalana, F., Knight, J., Russel, P., Omenetto, F., Efomov, A., and Taylor, A. (2003). *Nature*, 424, 511.
[23] Leon-Saval, S., Birks, T., Wadsworth, W., Russel, P., and Mason, M. (2004). *Opt. Express*, 12, 2864.
[24] Coen, S., Chau, A., Leonhardt, R., Harvey, J., Knight, J., Wadsworth, W., and Russell, P. (2002). *J. Opt. Soc. Am. B*, 19, 753.
[25] Wadsworth, W., Joly, N., Knight, J., Birks, T., Biancalana, F., and Russell, P. (2004). *Opt. Express*, 12, 299.
[26] Ravi, V., Kumar, K., George, A., Reeves, W., Knight, J., Russell, P., Omenetto, F., and Taylor, A. (2002). *Opt. Express*, 10, 1520.
[27] Gordon, J. P. (1986). *Opt. Lett.*, 11, 662
[28] Gordon, J. P. (1987). *Opt. Lett.*, 11, 659
[29] Solli, D., Ropers, C., Koonath, P., and Jalali, B. (2007). *Nature*, 450, 1054
[30] Monfared, Y., and Ponomarenko, S. (2017). *Opt. Express*, 25, 5941.
[31] Ankiewicz, A., Bokaeeyan, M., and Akhmediev, N. (2018). *J. Opt. Soc. Am. B*, 35, 899.
[32] Hasegawa, A., and Brinkman, W. (1980). *IEEE J. Quantum Electron.*, 16, 694.
[33] Tai, K., Hasegawa, A., and Tomita, A. (1986). *Phys. Rev. Lett.*, 56, 135.
[34] Mussot, A., Lantz, E., Maillotte, H., Sylvestre, T., Finot, C., and Pitois, S. (2004). *Opt. Express*, 12, 2838
[35] Biancalana, F., Skryabin, D., and Yulin, A. (2004). *Phys. Rev. E*, 70, 016615.
[36] Cumberland, B., Travers, J., Popov, S., and Taylor, J. (2008). *Opt. Express*, 16, 5954.
[37] Kudlinski, A., George, A., Knight, J., Travers, J., Rulkov, A., Popov, S., and Taylor, J. (2006). *Opt. Express*, 14, 5715.
[38] Kudlinski, A., and Mussot, A. (2008). *Opt. Lett.*, 33, 2407.
[39] Ortigosa-Blanch, A., Knight, J. C., and Russel, P. (2002). *J. Opt. Soc. Am. B*, 19, 2567.

[40] Geny, G., Lehtonen, M., Ludvigsen, H., Broeng, J., and Kaivola, M. (2002). *Opt. Express*, 10, 1083.
[41] Gorbach, A., Skyabin, D., Stone, J., and Knight, J. (2006). *Opt. Express*, 14, 9854.
[42] Kodama, Y., and Hasegawa, A. (1987). *IEEE J. Quant. Electron.*, 23, 510.
[43] Akhmediev, N., and Karlsson, M. (1995). *Phys. Rev. A*, 51, 2602.
[44] Husakou, A., and Herrmann, J. (2001). *Phys. Rev. Lett.*, 87, 203901.
[45] Nishizawa, N., and Goto, T. (2002). *Opt. Express*, 10, 1151.
[46] Gorbach, A., and Skryabin, D. (2007). *Nature Photonics*, 1, 653.
[47] Sakamaki, S., Nakao, M., Naganuma, M., and Izutsu, M. (2004). *IEEE J. Sel. Top. Quantum Electron.*, 10, 876.
[48] Genty, G., Lehtonen, M., and Ludvigsen, H. (2004). *Opt. Express*, 12, 4614.
[49] Frosz, M., Falk, P., and Bang, O. (2005). *Opt. Express*, 13, 6181.
[50] Judge, A., Bang, O., and Sterke, C. (2010). *J. Opt. Soc. Am. B*, 27, 2195.
[51] Roy, S., Bhadra, S., Saitoh, K., Koshiba, M., and Agrawal, G. (2011). *Opt. Express*, 19, 10443.
[52] Bellini, M., and Hänsch, T. (2000). *Opt. Lett.*, 25, 1049.
[53] Udem, T., Holzwarth, R., and Hänsch, T. (2002). *Nature*, 416, 233.
[54] Fercher, A., Drexler, W., Hitzenberger, C., and Lasser, T. (2003). *Rep. Progr. Phys.*, 66, 239.
[55] Lindfors, K., Kalkbrenner, T., Stoller, P., and Sandoghdar, V. (2004). *Phys. Rev. Lett.*, 93, 037401.
[56] Hundertmark, H., Wandt, D., Fallnich, C., Haverkamp, N., and Telle, H. (2004). *Opt. Express*, 12, 770.
[57] Takara, H., Ohara, T., Yamamoto, T., Masuda, H., Abe, M., Takahashi, H., and Morioka, T. (2005). *Electron. Lett.*, 41, 270.
[58] Humbert, G., Wadsworth, W., Leon-Saval, S., Knight, J., Birks, T., Russell, P., Lederer, M., Kopf, D., Wiesauer, K., Breuer, E., and Stifter, D. (2006). *Opt. Express*, 14, 1596
[59] Labruyère, A., Tonello, A., Couderc, V., Huss, G., and Leproux, P. (2012). *Opt. Fiber Technol.*, 18, 375.
[60] Hakala, T., Suomalainen, J., Kaasalainen, S., and Chen, Y. (2012). *Opt. Express*, 20, 7119.
[61] Kawagoe, H., Ishida, S., Aramaki, M., Sakakibara, Y., Omoda, E., Kataura, H., and Nishizawa, N. (2014). *Biomed. Opt. Express*, 5, 932.
[62] Sun, C., Ge, T., Li, S., An, N., and Wang, Z. (2016). *Appl. Opt.*, 55, 3746.
[63] Qi, X., Chen, S., Li, Z., Liu, T., Ou, Y., Wang, N., and Hou, J. (2018). *Opt. Lett.*, 43, 1019.
[64] Zhao, L., Li, Y., Guo, C., Zhang, H., Liu, Y., Yang, X., Wang, J., Jing, F., and Feng, G. (2018). *Opt. Commun.*, 425, 118.
[65] Tao, Y., Chen, S., and Xu, H. (2019). *Opt. Express*, 27, 26044.
[66] Stone, J., and Knight, J. (2008). *Opt. Express*, 16, 2670.
[67] Ghosh, D., Roy, S., Pal, M., Leproux, P., Viale, P., Tombelaine, V., and Bhadra, S. (2011). *J. Lightwave Technol.*, 29, 146.
[68] Chen, H., Chen, Z., Zhou, X., and Hou, J. (2013). *Laser Phys. Lett.*, 10, 085401.

[69] Wang, N., Cai, J., Qi, X., Chen, S., Yang, L., and Hou, J. (2018). *Opt. Express*, 26, 1689.

[70] Moller, U., Yu, Y., Kubat, I., Petersen, C., Gai, X., Brilland, L., Méchin, D., Caillaud, C., Troles, J., Luther-Davies, B., and Bang, O. (2015). *Opt. Express*, 23, 3282.

[71] Cimalla, P., Walther, J., Mittasch, M., and Koch, E. (2011). *J. Biomed. Opt.*, 16, 116020.

[72] Ringsted, T., Siesler, H., and S. Engelsen, S. (2017). *J. Cereal Sci.*, 75, 92.

[73] Petersen, C., Prtljaga, N., Farries, M., Ward, J., Napier, B., Lloyd, G., Nallala, J., Stone, N., and Bang, O. (2018). *Opt. Lett.*, 43, 999.

[74] Islam, M., Freeman, M., Peterson, L., Ke, K., Ifarraguerri, A., Bailey, C., Baxley, F., Wager, M., Absi, A., Leonard, J., Baker, H., and Rucci, M. (2016). *Appl. Opt.*, 55, 1584.

[75] Domachuk, P., Wolchover, N., Cronin-Golomb, M., Wang, A., George, A., Cordeiro, C., Knight, J., and Omenetto, F. (2008). *Opt. Express*, 16, 7161.

[76] Liao, M., Gao, W., Duan, Z., Yan, X., Suzuki, T., and Ohishi, Y. (2012). *Opt. Lett.*, 37, 2127.

[77] Qin, G., Yan, X., Kito, C., Liao, M., Chaudhari, C., Suzuki, T., and Ohishi, Y. (2009). *Appl. Phys. Lett.*, 95, 161103.

[78] Wang, F., Wang, K., Yao, C., Jia, Z., Wang, S., Wu, C., Qin, G., Ohishi, Y., and Qin, W. (2016). *Opt. Lett.*, 41, 634.

[79] Kubat, I., Agger, C., Møller, U., Seddon, A., Tang, Z., Sujecki, S., Benson, T., Furniss, D., Lamrini, S., Scholle, K., Fuhrberg, P., Napier, B., Farries, M., Ward, J., Moselund, P., and Bang, O. (2014). *Opt. Express*, 22, 19169.

[80] Saini, T., Kum, A., Zhao, Z., Wang, X., Dai, S., Pan, Z., Liu, S., Sun, L., Zhang, P., Liu, Z., Nie, Q., Shen, X., Wang, R., and Sinha, R. (2015). *J. Lightwave Technol.*, 33, 3914.

[81] Zhao, Z., Wang, X., Dai, S., Pan, Z., Liu, S., Sun, L., Zhang, P., Liu, Z., Nie, Q., Shen, X., and Wang, R. (2016). *Opt. Lett.*, 41, 5222.

[82] Dai, S., Wang, Y., Peng, X., Zhang, P., Wang, X., and Xu, Y. (2018). *Appl. Sci.*, 8, 707.

[83] Petersen, C., Møller, U., Kubat, I., Zhou, B., Dupont, S., Ramsay, J., Benson, T., Sujecki, S., Abdel-Moneim, N., Tang, Z., Furniss, D., Seddon, A., and Bang, O. (2014). *Nat. Photonics*, 8, 830.

[84] Zhang, B., Yu, Y., Zhai, C., Qi, S., Wang, Y., Yang, A., Gai, X., Wang, R., Yang, Z., Luther-Davies, B., and Xu, Y. (2016). *J. Am. Ceram. Soc.*, 99, 2565.

[85] Cheng, T., Nagasaka, K., Tuan, T., Xue, X., Matsumoto, M., Tezuka, H., Suzuki, T., and Ohishi, Y. (2016). *Opt. Lett.*, 41, 2117.

[86] Dudley, J. and Coen, S. (2002). *Opt. Lett.*, 27, 1180.

[87] Huang, C., Liao, M., Bi, W., Li, X., Hu, L., Zhang, L., Wang, L., Qin, G., Xue, T., Chen, D., and Gao, W. (2018). *Photon. Res.*, 6, 601.

[88] Diouf, M., Cherif, R., Ben Salem, A., and Waguem, A. (2017). *J. Mod. Opt.*, 64, 1335.

[89] Diouf, M., Salem, A., Cherif, R., Saghaei, H., and Wague, A. (2017) *Appl. Opt.*, 56, 163.

[90] Heidt, A. (2010). *J. Opt. Soc. Am. B*, 27, 550.
[91] Heidt, A., Hartung, A., Bosman, G., Krok, P., Rohwer, E., Schwoerer, H., and Bartelt, H. (2011). *Opt. Express*, 19, 3775.
[92] Hooper, L., Mosley, P., Muir, A., Wadsworth, W., and Knight, J. (2011). *Opt. Express*, 19, 4902.
[93] Zhang, N., Peng, X., Wang, Y., Dai, S., Yuan, Y., Su, J., Li, G., Zhang, P., Yang, P., and Wang, X. (2019). *Opt. Express*, 27, 10311.
[94] Stolen, R. H., Gordon, J. P., Tomlinson, W. J., and Haus, H. A. (1989) *J. Opt. Soc. Am. B*, 6, 1159.
[95] Blow, K. J., and Wood, D. (1989). *IEEE J. Quantum Electron.*, 25, 12, 2665.
[96] Lin, Q., and Agrawal, G. P. (2006). *Opt. Lett.*, 31, 21, 3086.
[97] Kodama, Y., and Hasegawa, A. (1987). *IEEE J. Quantum Electron.*, 23, 510.
[98] Chen, C. M., and Kelley, P. L. (2002). *J. Opt. Soc. Am. B*, 19, 196
[99] Dianov E. M., Nikonova Z. S., Prokhorov A. M., and Serkin, V. N. (1986). *Pis'ma Zh. Tekh. Fiz.* 12, 756 [*Sov. Tech. Phys. Lett.*, 12, 311].
[100] Husakou, A., and Herrmann, J. (2001). *Phys. Rev. Lett.*, 87, 203901.
[101] Skryabin, D. V., Luan F., Knight J. C., and Russell P. S. (2003). *Science*, 301, 1705
[102] Tsoy, E., and deSterke, C. M. (2006). *J. Opt. Soc. Am. B*, 23, 2425.
[103] Rodrigues, S., Facão, M., Latas, S., and Ferreira, M. (2013). *Photon. Nanostruct*, 11, 226.
[104] Rodrigues, S., Facão, M., and Ferreira, M. (2017). *J Nonl. Opt. Phys. Mat.*, 56, 1750049.

Chapter 5

A Highly Nonlinear Multi-Material Effectively Single-Mode Photonic Crystal Fiber for Wideband Parametric Sources in a Near-IR Window

Bhaskar Ojha[1]
Rohaan Deb[1,2]
Supratim Chatterjee[1]
and Partha Sona Maji[1,*]**, PhD**
[1]Department of Physics, Amity School of Applied Sciences,
Amity University, Kolkata, India
[2]Department of Physics, Bonn-Cologne Graduate School of Physics and Astronomy,
Universität Bonn, Bonn, Germany

Abstract

In this chapter, we detail realization of wide bandwidth with high gain parametric sources based on dispersion engineered highly nonlinear multi-material photonic crystal fiber (PCF). The multi-material PCF is composed of highly nonlinear As_2S_3 as the background material and borosilicate rods replace the first air-hole ring for a suitable refractive index profile towards a unique dispersion profile for high gain with a wider bandwidth parametric source. The PCF profile is engineered wisely towards a D characteristics having near-zero value and support small along with a +ve value of fourth order dispersion parameter (β_4)

* Corresponding Author's Email: parthamaji.1984@gmail.com.

In: Photonic Crystal Fibers (PCFs)
Editor: Mário F. S. Ferreira
ISBN: 979-8-88697-618-2

favouring the condition for very wide fiber optic parametric amplifier (FOPA) bandwidth. We investigated the dependency of pump power, fiber length and operating wavelength on parametric gain and bandwidth for fixed values of the other associated parameters. Each parameter gives rise to some unique findings of the amplifier performances. The operating wavelength in the anomalous and normal dispersion region plays a huge role on the peak gain and overall bandwidth of the amplifier. A judicious choice of fiber length and pump power is also required for a wide 3 dB bandwidth for applications in DWDM network system along with a required suppression of other unwanted nonlinear effects. Based on the numerical studies, we obtained few configurations which support a very wide 3dB FOPA bandwidth of more than 2000 nm in the near IR region. The presented approach for dispersion engineering and wide bandwidth will be very much helpful for the researchers in the field and will pave way new directions for future efficient sources in different wavelength window.

Keywords: photonic crystal fiber (PCF), four wave mixing, parametric amplifier, chalcogenide material, nonlinear optics

Introduction

FOPAs have been seen to attract a considerable amount of attention since they can provide broadband amplification covering both the telecommunication band as well as IR application [1, 2]. A FOPA is based on highly efficient four wave mixing (FWM), has been experimentally investigated since the late 1980s [3]. It offers a wide gain bandwidth, similar to a Raman amplifier, as well as having noise penalties lower than the 3 dB quantum-limit. [4]. The above achievement has been possible because of the recent development in the field of optical fiber communication due to the availability of good quality pump sources and fiber materials [1, 2]. In spite of all this, FOPAs' performance is limited due to the lower nonlinearity of silica based optical fibers. In this context, highly non-linear fibers material such as tellurite and chalcogenide-based optical fibers have gained high popularity as they were able to offer wide transmission windows having high nonlinearity [3, 5-17].

In this Chapter, highly non-linear chalcogenide-based fiber has been investigated towards a unique dispersion profile which suits the condition for high gain and wide bandwidth. We considered a PCF profile where the first air-hole ring is replaced with fabrication compatible Borosilicate with a refractive index (RI) larger than silica thereby creating a unique RI profile suitable for wider bandwidth having a higher gain. We investigated different profiles for our work having two ZDWs. The FOPA performance has been investigated around the second ZDW as a function of fiber length, pump power, and operating wavelength in this work. Our numerical investigation results a FOPA source having 3 dB bandwidth more than 1000 nm in the near-IR region. The investigated research is very new and the obtained bandwidth is extremely wide and the concept will be very helpful for various photonic devices in the coming future.

PCF Design and Analysis Method

In this work, a highly non-linear chalcogenide-based PCF with uniform air hole diameter has been proposed. The background material is As_2S_3 with the first air hole ring is replaced through borosilicate glass rods to have a favorable dispersion profile. The hybrid combination of As_2S_3 and borosilicate glass will be a feasible combination since the materials have compatible thermal properties along with the available well-matured fabrication techniques [18-20]. The schematic of the studied fiber is illustrated in Figure 1 where the black color presents the background material of As2S3, the white circles symbolize the air-hole rings and the grey circles correspond to the borosilicate glass rods respectively with pitch (Λ) is the distance between two neighboring air-holes.

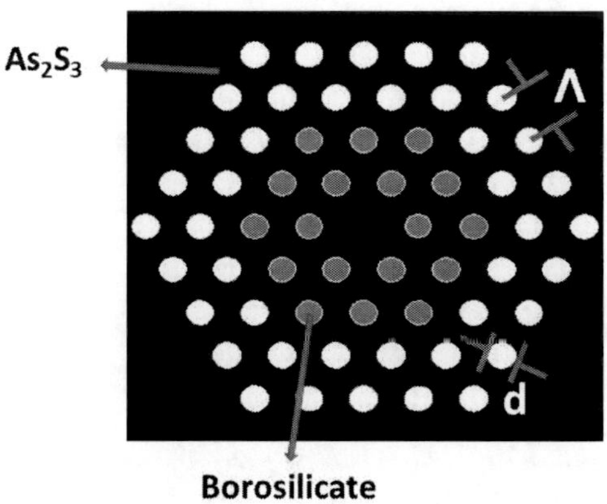

Figure 1. Schematic cross-section of the projected PCF structure with *d* as the air hole diameter and *Λ* as the hole-to-hole distance. The background material is AS_2S_3 with the first inner air hole ring is replaced with borosilicate rods.

For the ease of the fabrication process, the diameter of all the holes in the 3 rings has been kept constant. The mode profile has been evaluated based on CUDOS based MOF utilities which makes use of multipole method [21] whose validity and justification are well established [22, 23]. The dispersion parameter D is evaluated with Eq. (1)

$$D = (-\lambda/c)\, d^2\, \mathrm{Re}\left[n_{eff}\right]/d\lambda^2 \qquad (1)$$

where $\mathrm{Re}\left[n_{eff}\right]$ is the real part of effective indices obtained from simulation and c is the velocity of light in vacuum. The refractive index profile of the constituent material of the micro-structured fiber has been considered based on the standard Sellmeier's equation with C_1=1.898367, C_2=1.922297, C_3= 0.87651, C_4= 0.11887, C_5= 0.95699, λ_1=0.0225 μm, λ_2=0.0625 μm, λ_3=0.1225 μm, λ_4 =0.2025μm, λ_5=0.3025 for AS_2S_3 [15] and C_1=0.967, C_2=0.00511, C_3= 0.34938, λ_1=0.020452 μm, λ_2=107.9261 μm, λ_3=0.000233 μm for borosilicate [15].

Theory of Single Pump FOPA

As mentioned earlier, the theory of a single pump-based parametric amplifier is completely based on the FWM process which is generally a process in which energy is transferred from strong pump fields to signal and idler waves [1]. In short, it is a nonlinear process that contributes to the generation of new frequencies and has applications in the development of tunable lasers, wavelength convertible fiber optic parametric amplifiers (FOPAs) and multiwavelength fiber lasers [4]. In a single pumping scheme, two pump photons having the same frequency ω_p are converted into signal and idler photons with frequencies greater (ω_s) and lower (ω_i) than the pump photon respectively i.e., the signal photon has a lower wavelength in comparison to the idler photon. In short, they follow the following energy relation [2]:

$$2\omega_p = \omega_s + \omega_i \qquad (2)$$

The optical parametric gain G_s is given by Eq. (3) [2] as follows:

$$G_s = 1 + \left(\frac{\gamma P}{g}\right)^2 \sinh^2(gL) \tag{3}$$

where, γ is the effective nonlinear coefficient given by $\gamma = 2\pi n_2/(\lambda A_{eff})$ (n_2 is the non-linear refractive indices of the background material having values of 4.2×10^{-18} m^2/W for the present case of AS_2S_3 and A_{eff} is the effective area of the fundamental mode) and P is the input pump power with g as the parametric gain co-efficient given as follows

$$g = \sqrt{(\gamma P)^2 - (\Delta k/2)^2} \tag{4}$$

where, $\Delta k = 2\gamma P + \Delta k_L$, Δk_L is the linear phase mismatch term also mentioned as $\Delta\beta$. In order to realize the net amplification of the signal source, the parametric gain coefficient, g should be real. Therefore, acceptable values of $\Delta\beta$ for single pump case are limited to the range $-4\gamma P \le \Delta\beta \le 0$. Additionally, the FOPA gain is highest when Δk =0 or when $\Delta\beta = -2\gamma P$. It must be noted that the waveguide phase mismatch term Δk_W can be neglected for single-mode fibers [1]. The value of $\Delta\beta$ is known by the following relation [2]:

$$\Delta\beta = \left[2\sum_{m=1}^{\infty} \frac{1}{(2m)!} \beta_{2m} (\Delta\omega)^{2m}\right] \tag{5}$$

where $\Delta\omega$ is the frequency shift ($\Delta\omega = \omega_s - \omega_p = \omega_p - \omega_i$) for m=1 and 2, β_{2m} corresponds to β_2 and β_4 which are respectively the 2nd and 4th higher order dispersion (HOD) parameters, where $\beta_2 = -\frac{\lambda^2}{2\pi c} D$ with D is the dispersion parameter and β_4 is related to β_2 as $\beta_4 = \partial^2\beta_2/\partial w^2$.

Results and Discussions

In order to realize near-zero dispersion around near-IR wavelengths, unique dispersion management is required for the interplay between material and waveguide dispersion. The waveguide dispersion can be engineered via varying the air-hole diameter and Λ of the PCF structure along with suitable fabrication compatible multi-material composition. We have considered a Λ value of 1.75 μm for which the dispersion results for varying air-hole diameter "d" are presented in Figure 2 for a range of 0.85 μm to 1.25 μm. It can be evidently observed that with the increase of air-hole diameter, the D nature presents two ZDWs which go on getting separated away from each other at the same time the peak dispersion increases. These facts have some important consequences. Firstly, to achieve a broadband FOPA source the two ZDWs should be moderately separated such that the optimized gain profile together doesn't overlap much giving rise to an overall very wide spectrum. Secondly, the D value should remain nearer to zero such that phase mismatch remains well within the allowed limit $-4\gamma P \leq \Delta\beta \leq 0$ towards high and broadband FOPA bandwidth. Based on the above factors, we considered d=1.09 μm (for the above mentioned value of Λ) for our investigation for a parameter dependency of the FOPA gain and bandwidth. This should be noted that similar dependency took place for other air-hole diameter also. This is also to be noted that we tried to keep the diameter relatively bigger to ensure fabrication easiness especially in the case of multimaterial PCFs. At the same time, it should be noted that for d/Λ =0.63 (1.10/1.75) the fiber is not endlessly single-mode [24] for the whole wavelength, however, it has been observed that the propagation loss of the next higher-order mode to that of the fundamental mode is ~ 10^3 times which makes the fiber effectively single-mode for the wavelength of interest. Figure 3 illustrates the relevant dispersion nature for d=1.09 μm with the relevant 2nd order and 4th order HOD terms (β_2 and β_4 respectively) in the bottom and top inset. It is interesting to observe that the precise value of ZDWs for these two cases are correspondingly 2296 nm and 3576 nm. The dispersion profile is reasonably flat (leading towards possibility of reasonable wavelength tuning around the ZDWs) with the two ZDWs are approximately ~1300 nm (1280 nm exactly) separated from each other that may guide towards a very wide combined bandwidth. It is interesting to observe that around these two ZDWs, the value of β_4 is positive which is going to play a huge role in the gain spectrum since the sign and value of the 4th order HOD term plays a very important role in the spectrum of the FOPA gain and bandwidth [25, 26].

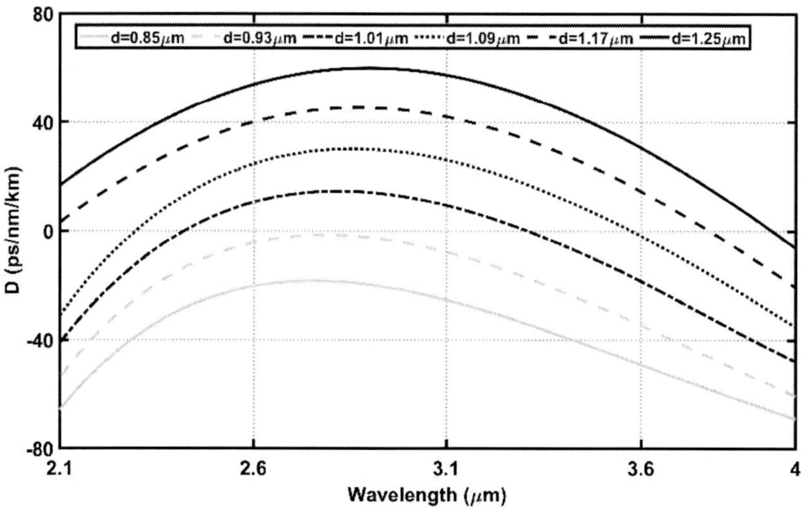

Figure 2. The calculated dispersion profiles of the considered PCF with different diameters including both the air-holes and borosilicate (from 0.90 to 1.25 μm) for Λ = 1.75 μm.

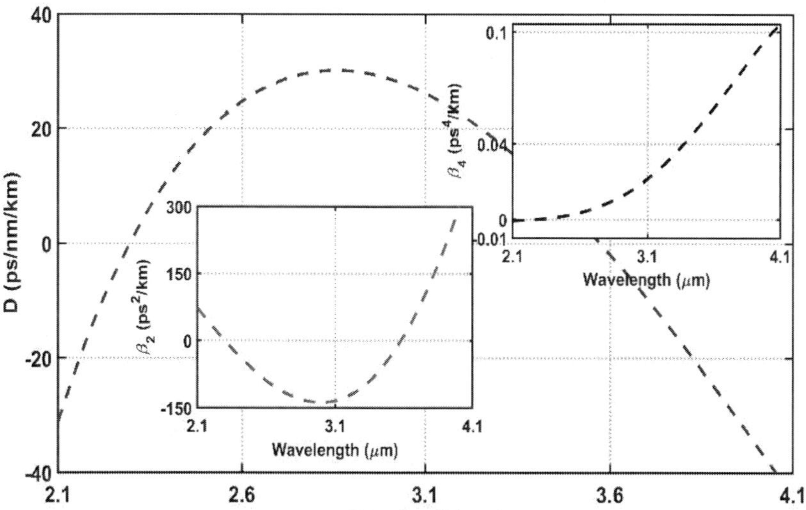

Figure 3. Dispersion profile with near-zero with flat nature for Λ = 1.75 μm d = 1.09 μm. The variation of 2nd order HOD (β_2) and 4th order HOD (β_4) parameters with wavelength are presented in the bottom and top inset respectively. This is interesting to note that β_4 is +ve around both the ZDWs.

This is to be noted that similar study has been carried for all the other D profiles also whose optimized profile is presented in the later section of the article. We now carry out a detailed investigation of the FOPA performances based on the obtained ZDWs.

We investigatied the FOPA performance around ZDW_1 which is 2296 nm in the following section.

FOPA Performance Around ZDW_1

Length Variation

An illustration of change of FOPA gain bandwidth with the variation of fiber length around ZDW_1 is shown in Figure 4 for a constant pump power of P=5W along for a pump wavelength of 2300*nm* for an anomalous pumping (while the ZDW_1 is 2296nm). The pump wavelength is considered in the anomalous region since a +ve β_4 gives broader bandwidth with high gain in this anomalous region only [12]. The figure demonstrates that there is an increase in peak gain with an overall decrease of the bandwidth of the amplifier with increasing fiber length for a constant pump power and operating wavelength. It should be well observed that although the peak gain increases with fiber length, the 3dB bandwidth which is a very important aspect of the DWDM system decreases. So, depending upon the fiber length, the peak gain and broadband spectrum can be easily achieved as per requirement. An additional important observation can be found from the above figure that we have additional peaks on both sides of the central gain spectrum centered around 1460 nm and 4600 nm on lower and higher wavelength part respectively. The above fact can be realized based on the interplay between β_2 and β_4 values towards total $\Delta\beta$. A +ve β_4 along with a -ve β_2 makes the total $\Delta\beta$ within $-4\gamma P \leq \Delta\beta \leq 0$ around these wavelengths to create a reasonable gain. Additionally, the slope of $\Delta\beta$ graph vs wavelength dictates the bandwidth of these peaks. A flatter slope will give wide bandwidth while a sharp slope will lead towards a very narrow source. These facts are explained in detail in the wavelength dependence upon the FOPA performance section. Additionally, due to the very high value of the nonlinear coefficient of the background material, the required length of the FOPA is kept smaller ensuring low propagation loss of the device.

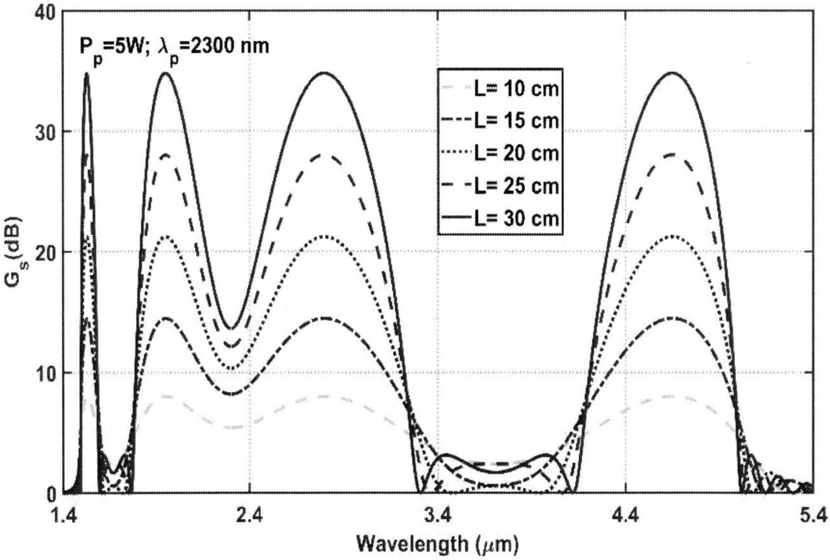

Figure 4. Gain and bandwidth variation with fiber length of the hybrid PCF for P=5W and pump wavelength of 2300 nm.

Power Variation

A variation of parametric gain bandwidth with the variation of pump power is presented in Figure 5 for a pump wavelength of 2300 nm and 20cm of fiber length. The figure demonstrates the fact that there is an increase in peak gain and bandwidth with the increase of the pump power; however, the 3dB bandwidth diminishes for the central bandwidth. Similar to Figure 4 additional peaks are well observed on either sides of the center spectrum with broader bandwidth with pump power increments. An increase in the pump power increases the spectrum window for $-4\gamma P \leq \Delta\beta \leq 0$ leading towards a wider spectrum for higher pump powers. A proper value of pump power required to be adjusted towards a uniform gain and bandwidth suitable for communication purposes. A high pump power can also induce other associated effects like an additional Raman spectrum [25-26]. We calculate the Raman threshold of the present fiber using relation $P_{TH} = 16A_{eff}/g_R L_{eff}$ [26]. Using Raman gain coefficient for As_2S_3 based chalcogenide material as 4.4×10^{-12} m/W [26] the calculated Raman threshold of the proposed fiber (considering the A_{eff} around pump wavelength to be 3.66 μm^2) is found to be ~66W for a fiber length of 20 cm. The value is much higher than the considered value of the pump power (10W) which signifies that the Raman scattering effect does not influence the FOPA gain spectrum.

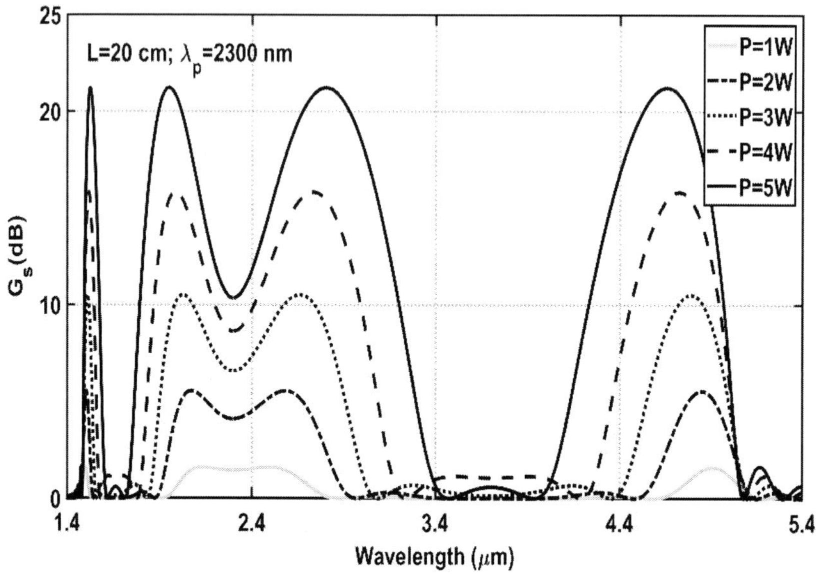

Figure 5. Gain and bandwidth variation with pump power for 20 cm of fiber at λp = 2300 nm.

Wavelength Variation

The operating pump wavelength plays a huge role in the overall behavior of FOPA. The FOPA gain spectrum and the associated phase-mismatch term $\Delta\beta$ for a variation of λ_p is presented in Figure 6(a) and 6(b) respectively for a pump power of 4W and a fiber length of 20 cm. Improved phase matching and consequent superior FOPA gain with wider bandwidth are achieved around smaller anomalous pumping (i.e., $\beta_2 \leq 0$) having a + β_4 [25]. This should be noted that the considered PCF structure is having a positive β_4 in the region of the ZDW_1 as shown on the top inset of Figure 3. Additionally, the optimized fiber has a relatively flatter dispersion profile close to zero value; hence the pumping condition can be tuned further from the ZDW towards the anomalous region for an improved phase matching and consequent wider bandwidth with an adequate FOPA gain.

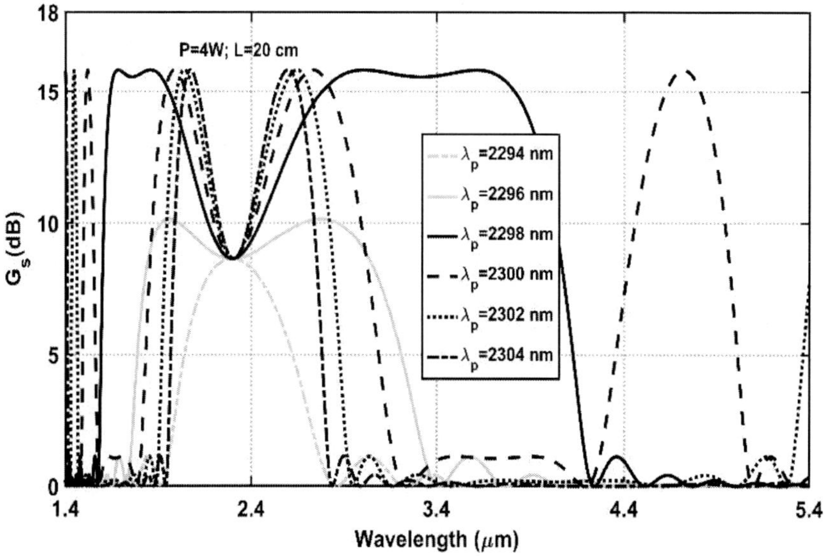

(a)

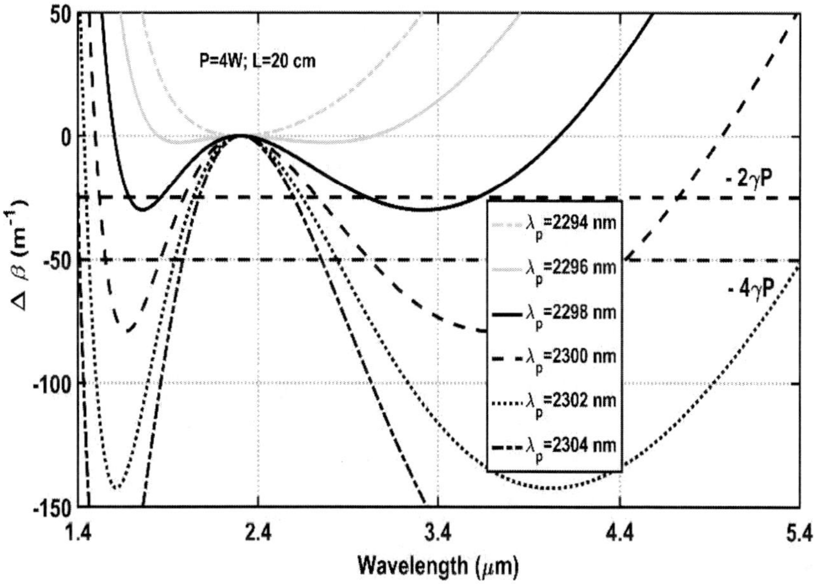

(b)

Figure 6. (a) λ_p effect on gain spectrum for P=4 W and 20 cm fiber length in the region of ZDW_1; (b) (Δβ) deviation analogous to (a).

We started our investigation of pump wavelength dependency with a normal pumping ($\beta 2 > 0$) for $\lambda p=2994$ nm (with dash light black line). For this wavelength, the value of $\Delta\beta$ is well above the range for a high and broadband gain for $-4\gamma P \leq \Delta\beta \leq 0$ .resulting in lower gain and narrow bandwidth. We move towards another operating wavelength of $\lambda p=2296$ nm which is very close to the ZDW1. This may be noted that the exact ZDW1 value is in between 2296 nm and 2297 nm with close proximity towards the former one. For this operating wavelength (presented by solid light black line) $\Delta\beta$ is within the gain condition only for a small wavelength region resulting in lower and narrow gain as shown in Figure 6(a). As we move towards the operating wavelength of 2998nm we can observe that the $\Delta\beta$ is within the gain condition for a very wide range resulting in a very broadband spectrum on either sides of the pump. Additionally, since $\Delta\beta$ is around $-2\gamma P$ for a reasonable wider wavelength this corresponds to an almost flat spectrum with little fluctuations in gain profile as can be clearly seen in Figure 6(a) solid black line. As we tune the wavelength further for $\lambda p=3000$nm, the value of $\Delta\beta$ goes outside the allowed limit resulting in a narrow bandwidth. However, for a wider wavelength choice on both sides of the central gain region, the value of $\Delta\beta$ again enters within the allowed limit resulting in narrow peaks on both sides. A sharper slope of the left window causes a narrow peak while a relatively horizontal variation causes a much broader bandwidth on the long-wavelength part of the window shown with big dash black line. Moving further for $\lambda p=3000$ nm, $\lambda p=3002$ nm and $\lambda p=3004$ nm the central bandwidth goes on narrowing as the fewer amount of regions within $-4\gamma P \leq \Delta\beta \leq 0$ .however the new peaks on the two sides keep on moving towards wider regions both on lower and higher wavelength parts. So, dynamic amplifier performance for tuning of wavelength is possible as per the requirements. So, with the tuning of the operating wavelength new parametric sources much away from the pump can be achieved thereby creating an opportunity of achieving sources that otherwise is not possible to achieve with other methods.

Optimized Gain Profile Around ZDW_1

Based on the previous dependency of the individual parameters on the overall gain and bandwidth, the parametric gain profile has been optimized and the same is demonstrated in Figure 7 with the corresponding phase mismatch term presented in the bottom inset of the same figure. We obtained a 3 dB bandwidth of ~2500 nm with a peak gain of 7.2 dB with a pump power of 3 W for a fiber length of 16 cm. The dips and peaks of the gain continuum can easily be understood from the phase-mismatch curve with a net gain lying between $-4\gamma P \leq \Delta\beta \leq 0$ with maximum value at $\Delta\beta = -2\gamma P$. It is very interesting to note that a very wide range of spectrum of the higher wavelength range of the pump is having small gain fluctuations. A lesser variation in power in a broadband source is very useful since the smaller power variation between those wavelengths given a very wide smooth spectrum for DWDM network communications where the wavelength separation is very narrow.

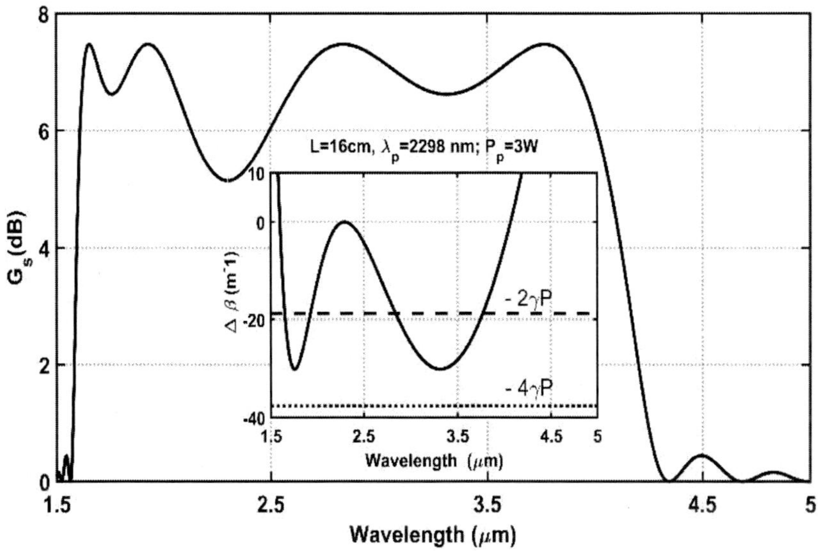

Figure 7. Optimized gain profile with a bandwidth of ~2500 nm for a fiber length of 16*cm* and pump power of 3W for a pump wavelength of 2298 nm and corresponding phase mismatch variation in the bottom inset.

Conclusion and Discussion

A novel technique for controlling dispersion profile in index-guiding PCFs is employed to design a highly nonlinear near-zero flat dispersion PCF for mid-IR applications. This design consists of three rings in the cladding with the innermost ring filled with borosilicate glass with an As_2S_3 background. Hence this design presents a RI-profile that provides not only a large nonlinear coefficient but also a flat dispersion slope with low propagation loss. We investigated different D profile around the near IR region such that the profile suits a very wide bandwidth FOPA spectrum. With a pitch value of 1.75 μm, we vary the diameters (including that of air-holes and borosilicate) towards investigating the FOPA characteristics around these profiles. We detailed the case for d = 1.09 μm and shows the detailed pump power, fiber length and operating wavelength (both normal and anomalous) effect upon gain profile. An anomalous pumping with a +ve β_4 corresponds to a high gain and wide bandwidth. After studying the outcomes for the different profiles, we optimized the structure having Λ = 1.75 μm and d = 1.09 μm with a 3 dB bandwidth source of more than 2500 nm around ZDW_1. The presented methodology will be highly beneficial towards generating useful sources with similar or even more suitable combinations in the near IR region based on the multi-material PCF geometry.

It would be very exciting to undertake fabrication of the considered present chalcogenide fibers design. The issue of wide glass transition temperature of the two glass combinations considered in the proposed design modeling can be solved by appropriate choice of fiber drawing temperature and concentration of the dopants like boron [19, 27]. Additionally, the presented material combination has been suggested by different research groups [19, 28, 29] for a futuristic exploration. Also to fabricate various chalcogenide PCFs, we already have well-matured fabrication technologies [10-11, 16-17, 30-34]. We believe that with the advancement of technologies similar fiber configurations will be easily drawn in near future. It is interesting to acknowledge that, once the proposed design (fiber profile) is accepted, fiber fabricators/manufacturer can always look for other appropriate glass combinations to solve fabrication issues using our proposed design as the initial design platform and if necessary fine tune the design parameters [19].

References

[1] Agrawal G. P., *Fiber-Optic Communication Systems*, 3rd ed. (Wiley, New York, (2002).

[2] Agrawal G. P., *Nonlinear Fiber Optics*, Optics and Photonics Series, 4th ed. (Academic Press 2007).

[3] Sanghera S., I. D. Aggarwal, L. B. Shaw, L. E. Busse, P. Thielen, V. Nguyen, P. Pureza, S. Bayya, and F. Kung, "Applications of chalcogenide glass optical fibers at NRL," *J. Optoelectron. Adv. Mater.*, 3, 627–640 (2001).

[4] Azura A. N., A. Harith and H. S. Wadi, *"Four Wave Mixing in a Highly Nonlinear Media,"* Lambert Academic Publishing (2014) ISBN: 978-3-659-34932-4.

[5] Camerlingo A., X. Feng, F. Poletti, G. M. Ponzo, F. Parmigiani, P. Horak, M. N. Petrovich, P. Petropoulos, W. H. Loh, and D. J. Richardson, "Highly-stable monolithic femtosecond Yb-fiber laser system based on photonic crystal fibers," *Opt. Express*, 18, 15747–15756 (2010).

[6] Martynkien T., D. Pysz, R. Stępień, and R. Buczyn′ ski, "All-solid microstructured fiber with flat normal chromatic dispersion," *Opt. Lett*. 39, 2342–2345 (2014).

[7] Saitoh K. and M. Koshiba, "Highly nonlinear dispersion-flattened photonic crystal fibers for supercontinuum generation in a telecommunication window," *Opt. Express,* 12, 2027–2032 (2004).

[8] Maji P. S. and P. R. Chaudhuri, "A new design for all-normal near zero dispersion photonic crystal fiber with selective liquid infiltration for broadband supercontinuum generation at 1.55 μm," *J. Photon*. 2014, 728592 (2014).

[9] Yamashita T. and Y. Ohishi, "Cooperative energy transfer between Tb3+ and Yb3+ ions co-doped in borosilicate glass," *J. Non-Cryst. Solids*, 354, 1883–1890 (2008).

[10] Tao G., S. Shabahang, E. H. Banaei, J. J. Kaufman, and A. F. Abouraddy, "Multimaterial preform coextrusion for robust chalcogenide optical fibers and tapers," *Opt. Lett*., 37, 2751–2753 (2012).

[11] Tao G., A. F. Abouraddy, and A. M. Stolyarov, "Multimaterial fibres," *Int. J. Appl. Glass Sci*., 3, 349–368 (2012).

[12] Deb R. and P. S. Maji, "Effect of HOD parameters on gain and bandwidth of parametric amplifiers for near-zero ultra-flat dispersion profile," *Optik: International Journal for Light and Electron Optics*, 195, 163158, (2019).

[13] Da N., L. Wondraczek, M. A. Schmidt, N. Granzow, and P. St. J. Russell, "High index-contrast all-solid photonic crystal fibers by pressure- assisted melt infiltration of silica matrices," *J. Non-Cryst. Solids*, 356, 1829–1836 (2010).

[14] Granzow N., P. Uebel, M. A. Schmidt, A. S. Tverjanovich, L. Wondraczek, and P. St. J. Russell, "Bandgap guidance in hybrid chalcogenide-silica photonic crystal fibers," *Opt. Lett*., 36, 2432– 2434 (2011).

[15] Chaudhari C., T. Suzuki, and Y. Ohishi, "Design of zero chromatic dispersion chalcogenide As_2S_3 glass nanofibers," *J. Lightwave Technol*., 27, 2095–2099 (2009).

[16] Eggleton B. J., B. L. Davies, and K. Richardson, "Chalcogenide photonics," *Nat. Photonics,* 5, 141–148 (2011).

[17] Hewak D. W., "The promise of chalcogenides," *Nat. Photonics*, 5, 474 (2011).

[18] Maji P. S. and P. Roy Chaudhuri, "Design of all-normal dispersion based on multi-material photonic crystal fiber in IR region for broadband supercontinuum generation," *Appl. Opt*. 54, 4042-4048 (2015).

[19] Barh A., S. Ghosh, R. K. Varshney, and B. P. Pal, "An efficient broad-band mid-wave IR fiber optic light source: design and performance simulation," *Opt. Express* 21, 9547-9555 (2013).

[20] Chaudhari, C., Suzuki, T., and Y. Ohishi, "Design of Zero Chromatic Dispersion Chalcogenide As_2S_3 Glass Nanofibers," *J. Lightwave Technol*. 27, 2095-2099 (2009).

[21] CUDOS MOF utilities, http://www.physics.usyd.edu.au/cudos/mofsoftware/.

[22] White T. P., B. T. Kuhlmey, R. C. McPhedran, D. Maystre, G. Renversez, C. M. de Sterke, and L. C. Botten, "Multipole method for microstructured optical fibers. I. Formulation," *J. Opt. Soc. Am. B*, 19, 2322–2330 (2002).

[23] Kuhlmey B. T., T. P. White, R. C. McPhedran, D. Maystre, G. Renversez, C. M. de Sterke, and L. C. Botten, "Multipole method for microstructured optical fibers. II. Implementation and results," *J. Opt. Soc. Am. B.*, 19, 2331–2340 (2002).

[24] Boris T. Kuhlmey, Ross C. McPhedran, C. Martijn de Sterke, Peter A. Robinson, Gilles Renversez, and Daniel Maystre, "Microstructured optical fibers: where's the edge?," *Opt. Express* 10, 1285-1290 (2002).

[25] Maji P. S. and P. Roy Chaudhuri, "Gain and bandwidth investigation in a near zero ultra-flat dispersion PCF for optical parametric amplification around the communication wavelength," *App. Opt.*, 54. 3263-3272 (2015).

[26] Maji P. S. and P. Roy Chaudhuri, "A new design methodology of obtaining wide band high gain broadband parametric source for IR wavelength applications," *Journal of Applied Physics (AIP)*, 119, 113102 (2016).

[27] Yamashita T. and Y. Ohishi. Cooperative energy transfer between Tb^{3+} and Yb^{3+} ions co-doped in borosilicate glass. *J. Non-Cryst. Solids*. 2008; 354(17): 1883–1890.

[28] Chaudhari C., T. Suzuki and Y. Ohishi. Design of Zero Chromatic Dispersion Chalcogenide As_2S_3 Glass Nanofibers. *J. Lightwave Technol*. 2009; 27: 2095-2099.

[29] Kalraa S., Vyas S., Tiwari M., Ismail Y., Yupapin P., Ali J. & Singh G. Investigation of As_2S_3-borosilicate chalcogenide glass-based dispersion-engineerd photonic crystal fibre for broadband supercontinuum generation in the mid-IR region. *J. Mod. Opt*. 2020; 67(10):920–926.

[30] Sanghera S., I. D. Aggarwal, L. B. Shaw, L. E. Busse, P. Thielen, V. Nguyen, P. Pureza, S. Bayya, and F. Kung. Applications of chalcogenide glass optical fibers at NRL. *J. Optoelectron. Adv. Mater*. 2001; 3: 627–640.

[31] El-Amraoui M., G. Gadret, J. C. Jules, J. Fatome, C. Fortier, F. Désévédavy, I. Skripatchev, Y. Messaddeq, J. Troles, L. Brilland, W. Gao, T. Suzuki, Y. Ohishi, and F. Smektala. Microstructured chalcogenide optical fibers from As_2S_3 glass: towards new IR broadband sources. *Opt. Express*. 2010; 18: 26655-26665.

[32] Sanghera J., C. Florea, L. Busse, B. Shaw, F. Miklos, and I. Aggarwal. Reduced Fresnel losses in chalcogenide fibers by using anti-reflective surface structures on fiber end faces. *Opt. Express*. 2010; 18: 26760-26768.

[33] Quentin C., B. Laurent, H. Patrick, N. T. Nam, C. Thierry, R. Gilles, M. Achille, F. Julien, S. Frédéric, P. Thierry, O. Hervé, S. Jean-Christophe, and T. Johann. Fabrication of low losses chalcogenide photonic crystal fibers by molding process. *Proc. SPIE*. 2010; 7598,75980O.

[34] Poletti F., V. Finazzi, T. M. Monro, N. G. R. Broderick, V. Tse, and D. J. Richardson. Inverse design and fabrication tolerances of ultra-flattened dispersion holey fibers. *Opt. Express*. 2005; 13: 3728-3736.

Biographical Sketches

Name: Bhaskar Ojhà

Affiliation: HNBGU Srinagar, Uttrakhand

Education: B.Sc (H)- Physics

Business Address: NA

Research and Professional Experience: 1 year

Professional Appointments: NA

Honors: NA

Publications from the Last 3 Years: 0

Name: Rohaan Deb

Affiliation: Amity University Kolkata and Department of Physics, Bonn-Cologne Graduate School of Physics and Astronomy, Universität Bonn, Germany

Education: BSc

Business Address: Major Arterial Road(South-East), AA II, Rajarhat, Newtown, West Bengal 700135

Research and Professional Experience: 3

Professional Appointments: NA

Honors: NA

Publications from the Last 3 Years: 4 publications

Name: Partha Sona Maji

Affiliation: Amity University Kolkata, India

Education: PhD

Business Address: Major Arterial Road (South-East), AA II, Rajarhat, Newtown, West Bengal, India

Research and Professional Experience: 13 yrs

Professional Appointments: Post-Doctoral Fellow, National Post-Doctoral Fellow

Honors: N/A

Publications from the Last 3 Years: 6 publications

Chapter 6

Photonic Crystal Fibers for Terahertz Applications

Vikas Kumar
Harijyoti Mandal
and Sunil Kumar*
Department of Physics, Indian Institute of Technology Delhi,
Hauz Khas, New Delhi, India

Abstract

Photonic crystal fibers (PCFs) have, in recent years, regenerated significant interest due to progressive ease of synthesis and their use for specific purposes at terahertz (micro-to-millimeter) frequencies (wavelengths). Based on polymeric host materials, such microstructured fibers have been studied to show their potential applications in terahertz (THz) generation by nonlinear mixing, polarization splitters, and other device applications. The efficiency for THz generation in silicon-based microstructured fibers is comparatively much smaller. Polymers-based microstructured fibers possess structural flexibility and tailored dispersive factors and have the desired response in a broadband frequency range. Low-loss polymers such as teflon, polyethylene, polycarbonate, polystyrene, t-surupica, and topaz are considered for modified total internal reflection and photonic bandgap (PBG) engineering for controlled THz wave generation and propagation. In this chapter, we have described THz PCF structures' design in different lattices and configurations having different basis choices and near-core

* Corresponding Author's Email: kumarsunil@physics.iitd.ac.in.

In: Photonic Crystal Fibers (PCFs)
Editor: Mário F. S. Ferreira
ISBN: 979-8-88697-618-2

air-holes-based topology for achieving tailored dispersion, high birefringence, large mode area, and low bending and propagation losses. Different methods, i.e., the full vector plane wave expansion method, finite element method, and beam propagation method, are applied to establish the PCFs' modal field and THz wave propagation characteristics. Generation of narrow and broadband THz radiation, nonlinear wave mixing, and polarization splitting with a high extinction ratio achievable at a short device length have been demonstrated using these characteristics.

Keywords: photonic crystal fibers, nonlinear mixing, THz polarization splitters, modified total internal reflection, THz sources

Introduction

Terahertz (THz) electromagnetic waves are used in many applications, including imaging, sensing, spectroscopy, and communication (Pawar, 2013; Hangyo, 2015; Hafez, 2016; Dragoman, 2004; Song, 2011; Angelo, 2014; Kozlov, 2015; Jazbinsek, 2019; Yang, 2016; Xie, 2014; Bogue, 2018; Mittleman, 1999). The corresponding frequencies are typically in the range of 0.1-10 THz, i.e., midway between microwaves and infrared waves in the electromagnetic spectrum. This band of frequencies was termed as THz gap for a long time due to the unavailability of proper sources and detectors (Hangyo, 2015; Hafez, 2016; Song, 2011) but subsequently, a tremendous amount of research and development work on various THz devices has taken place in the last couple of decades, and hence, there is no THz gap as such anymore (Suizu, 2007; Zhou 2011; Qasymeh, 2012; Barh, 2015; Han, 2002; Islam, 2018; Li, 2013; Khaleque, 2015; Wang, 2018; Liu, 2019; Yan, 2019; Tang, 2018; Markov, 2013; Mazhorova, 2011). THz radiation is non-ionizing because of low photon energy; hence, it is appropriate for biological scanning and medical diagnosis (Pawar, 2013; Bogue, 2018; Mittleman, 1999). The large penetration depth for nonmetallic substances and the chemical specificity for various hazardous and explosive chemicals make this frequency band appropriate for security applications (Pawar, 2013; Hangyo, 2015; Xie, 2014). Larger than microwave frequencies provide better resolution in imaging applications (Bogue, 2018). Suitable waveguides for THz radiation, like those in the optical and near-infrared domain, can enhance the application prospects many folds (13-20). In this regard, photonic crystal fibers (PCFs) and microstructured fibers have provided opportunities to explore their use for not only the loss-less guidance and compact devices size but also many other device applications by taking advantage of the fact that various properties such as dispersion, propagation loss, birefringence, nonlinearity, etc. can be engineered by optimizing the design (Suizu, 2007; Zhou 2011; Qasymeh, 2012; Barh, 2015; Han, 2002; Islam, 2018; Li, 2013; Khaleque, 2015; Wang, 2018; Liu, 2019; Yan, 2019; Tang, 2018; Markov, 2013; Mazhorova, 2011). The optimal design of the PCF structures by applying different geometrical features, configurations, and materials opens new avenues for varieties of PCFs-based THz applications. Waveguides have done wonders in the optical frequency region; however, the same is lacking by miles in the THz frequency range, and it is slowly picking up by the resources available now for use by many researchers and engineers (Khaleque, 2015; Wang, 2018; Liu, 2019; Yan, 2019; Tang, 2018; Markov, 2013; Mazhorova, 2011; Chen, 2007; Hassani, 2008; Kumar, Frontier of Optics, 2020; Cosentino, 2016; Kao, 2017; Abina, 2016; Hazhorova, 2012; Exter, 1989; Blake, 2008).

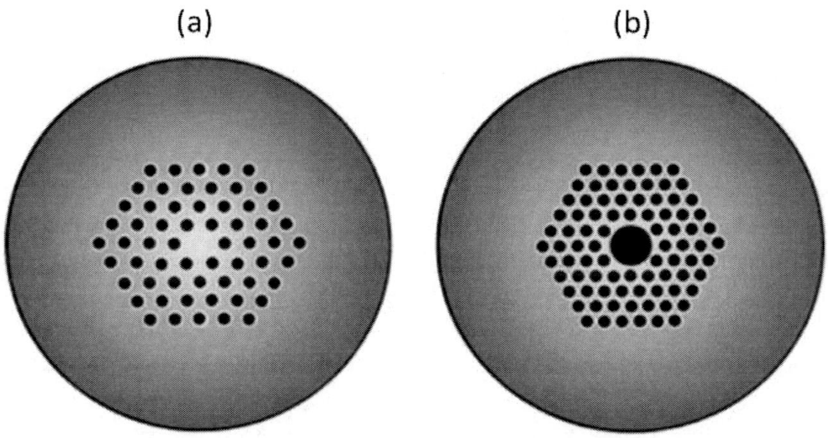

Figure 1. Two schemes of photonic crystal fibers. (a) Solid core index guided fibers and (b) hollow-core bandgap guided fibers. The black regions in both types of structures represent air holes.

PCFs can guide electromagnetic waves with desirable propagation characteristics compared to step-index fibers. Photonic crystal index guiding fibers (PCIGFs) are fibers consisting of a solid core and structured cladding that may consist of a periodic lattice of inclusions (air holes, generally) whose refractive index is lower than the host (background). One such possibility is shown in Figure 1(a). It causes a lower average refractive index for the cladding compared to the core refractive index, and hence, modified total internal reflection (MTIR) takes place (Sharma, 2013).

The lattices and bases in the microstructured region offer possibilities to improve certain performances of the THz PCIGFs. In the PBG fibers, a periodic repetition of dielectric and air hole structure in the cladding leads to bandgap in their photonic dispersion structure. The core consists of an appropriate size air hole; hence, PBG fibers are also known as photonic crystal hollow-core fibers, as illustrated in Figure 1(b), where the cross-sectional view is shown. An electromagnetic wave is guided through the fiber if its frequency vs. transverse wave vector point falls in the bandgap region. The core can be of a lower refractive index material also than the host material. Typically, the hollow core diameter is approximately 3 times larger than the pitch for the air holes in the cladding. Suitable alterations in the basis of the hole-lattice and the host material itself can help improve PBG spectral widths, and frequency shifts can be tuned (Kumar, Optics and Photonics: Theory and Computational Techniques Workshop and IEEE-WRAP, 2018, 2019).

This chapter has three main sections. The first provides basic details starting from suitable materials and microstructured configurations in the core/cladding regions to achieve desired characteristics, particularly the photonic band gap, tailored dispersion, and high birefringence. Terahertz polarization splitting in the compact size of the PCFs has been presented in the subsequent section, where two schemes of gradience in the material inclusions in the dual-core have been studied to achieve specific goals. The last section is devoted to generating THz radiation by nonlinear wave mixing in the PCFs. The focus has been kept on two points, narrow band with high power and broadband generation. Suitable waveguides have been designed by incorporating transparent materials for THz guidance in the PCFs and to achieve appropriate Rayleigh range, power transmission, etc., for specific applications. Polymers such as polyethylene, Teflon, polypropylene, polymethylmethacrylate, and polystyrene were considered for our PCF structures as they offer very low loss in the THz frequency range (Angelo, 2014).

THz Photonic Crystal Fibers

As mentioned before, photonic crystal fibers and microstructured fibers contain periodic lattice(s) of air holes or other material inclusions in a host material, either in the core or the clad. They can be classified into two general categories, (i) modified total internal reflection-based index guiding fibers (PCIGFs) (Sharma, 2013) and (ii) photonic band gap-based fibers (PBGFs) (Bouwmans, 2003), where the latter type is hollow-core fibers. Both possess characteristics that make them suitable for applications such as THz sources and various other devices (Russell, 2006; Barh, 2016). Though desired properties can be obtained by alteration of materials in the core region of Bragg fibers, formation of suspended core and anti-resonance reflecting optical wave guidance (ARROW) in Kagome-based fibers, but still there are several limitations (Healy, 2011). Bragg fibers offer limited tailoring due to the restriction to the hexagonal lattice of circular inclusions. Suspended core and Kagome PCF structures are fragile.

Traditionally, metallic THz waveguides have been studied, which are the scaled-down version of the microwave waveguides consisting of wires, transmission lines, coaxial cables, parallel plates, slots, or hollow tubes. However, propagation at THz frequencies is much affected by severe metallic losses, parasitic capacitances, and inductances, which also cause unwanted resonances (Gallot, 2000; Kozlov, 2015; Barh, 2016) and hence limit the operational bandwidth in any application. Silicon core step-index fibers were reported for broadband applications in the infrared and single-mode propagation of THz waves in conventional silica PCF only in a highly limited frequency range (Sørgård, 2020). Hollow-core PBG fibers with host materials silica and chalcogenides have also been reported for THz propagation (Pakarzadeh, 2019). Due to high loss and high dispersion at moderate and high THz frequencies (>0.5 THz), silicon, silica, or chalcogenides cannot be widely used for THz device applications (Barh, 2016; Sørgård, 2020). On the other hand, most of the polymers are low dispersive in a broadband THz frequency range. Polyethylene PCF with clad consisting of air holes in a hexagonal pattern (PBG fiber) were studied for THz waveguiding and group velocity dispersion (Han, 2002), however, they cannot help in dispersion tunability. Significantly large transparency could be achieved in a suspended core fiber using polyethylene as the host material; however, the presence of the struts not only limits it to use in a narrow frequency range of ~0.2-0.5 THz but also makes it fragile (Rozé, 2011). Several other designs of PCFs have also been reported to achieve either low bending-loss (Agarawal, 2012) or birefringence (Yang, 2018; Wu, 2018) but turned out to be difficult to fabricate (Pysz, 2016).

The polymer-based PCFs can be categorized as (i) hollow-core fibers, (ii) solid-core fibers, and (iii) porous core fibers. The guiding mechanisms for each of these are different. One can design special configurations to achieve either low-loss THz wave transmission or tailorable characteristics for specific THz devices. This chapter discusses various designs of suitable configurations in the periodic structures of polymers-based THz PCFs. The required wave propagation is achieved by changing the unit cell (basis) structure, its arrangement (lattice), and several types of materials suitable at the THz range. The use of transparent materials for proper THz guidance is necessary for which, polymers such as polyethylene, Teflon, polypropylene, polymethylmethacrylate, and polystyrene, have been considered (Angelo, 2014) and the stack and draw method is used to fabricate these PCFs (Pysz, 2016). For numerous applications, low dispersion, high birefringence, low bending and propagation losses, suitable modal area, high bandwidth, and single-mode guidance are desirable in the THz PCFs. Parameters including the effective indices (neff), birefringence (B), dispersion (D), effective mode areas (Aeff), nonlinearity, and numerical aperture (NA) (41) are tailored to meet specific needs. This has been achieved effectively by varying the lattice and basis structure in the core and the cladding regions. We have also discussed the limitations of conventional THz PCFs and ways to circumvent them by choosing the above schemes of microstructures in the core and cladding.

Photonic Bandgap

The design and properties of polyethylene (PE) based PBG fiber are presented here. The PCF consists of a hexagonal periodic lattice of air holes in the cladding with pitch (distance between the consecutive holes), $\Lambda = 300\mu m$, hole diameter, $d = 0.94\Lambda$, and radius of the core, = 1.42Λ. Figure 2 shows (a) the unit cell, (b) the Brillouin diagram in the Fourier space, (c) the supercell structure, and (d) the 3D wavevectors having k_z as the propagation vector. The photonic band diagram for normalized longitudinal propagation ($k_z = 10$) is presented in Figure 2(e), where a bandgap in it is clearly visible. Figure 2(f) describes the intersection method for determining the THz bandwidth of operation. Supercells are the basic structures consisting of all unit cells within the core of the PBG fiber under consideration. In the plane wave expansion method (PWEM) (Sukhoivanov, 2009; Angelo, 2014), numerous Bloch waves are considered in the governing Helmholtz equation and then a matrix of master equations for each k-vector is formed. Solving this matrix for the Eigenvalues provides the photonic band diagram consisting of the dispersion (ω, k) curves.

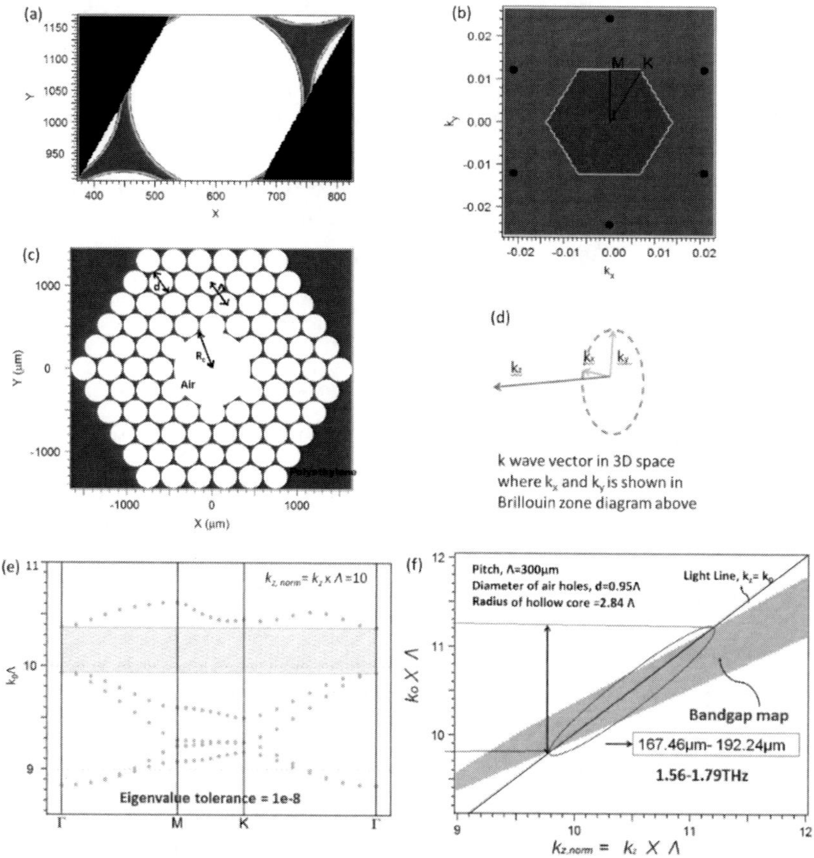

Figure 2. Polyethylene (PE) based PBG fiber and its properties. (a) Real space unit cell, (b) virtual space first Brillouin zone, (c) cross-section of the entire PBG fiber, (d) 3D visualization of the k-space for clarity, (e) photonic band structure for normalized longitudinal propagation, $k_z = 10$. The shaded region represents the bandgap. (f) Bandgap map at various k_z. d is the diameter of the air holes and Λ is the pitch.

The plane-wave expansion method was used to calculate band diagrams in all three dimensions. For the PCF structure shown in Figure 2(c), we obtained the corresponding band diagram as shown in Figure 2(e) consisting of all the k-eigenvalues of the matrix of master equations in the plane-wave expansion method. In the calculation, the eigenvalue tolerance was 10^{-8}, which is good enough for computation. For different polymers as host materials for the PBG fiber design as shown in Figure 2(c), where the corresponding refractive indices are taken from ref. (Angelo, 2014), the bandgap values for each longitudinal propagation vector, k_z are calculated. Considering 100 steps of these k_z values, the corresponding bandgap values are plotted in a single diagram, namely the bandgap map as shown in Figure 2(f) for polyethylene. The light-line is $k_z = k_0$, k_0 being the free space wave vector. An intersection of the light-line with the bandgap map provides the frequency bandwidth of the waves that can be guided through the hollow-core PBG fibers. By applying the same method to hollow-core PCFs of other polymers, a summary of the THz bandwidth for low-loss guiding through the PCFs is presented in Table 1 below.

Table 1. THz-bandwidth with low-loss propagation in hollow-core PBG fibers in various host polymers

Teflon (TPF)	1.65-1.95 THz (181.1-153.7 μm)
Polyethylene (PE)	1.56-1.79 THz (192.2-167.5 μm)
Polymethyl methacrylate (PMMA)	1.41-1.60 THz (213.2-188.3 μm)
Polycarbonate (PC)	1.35-1.48 THz (221.8-201.3 μm)
Cyclic olefin copolymer (COC)	1.53-1.75 THz (196.0-171.4 μm)

The total propagation loss (*PL*) due to material and, the confinement loss and the dispersion properties of PE-based PBG fiber (shown in Figure 2(c)), have been evaluated using the finite element method (FEM) (Rahman, 2013) and the corresponding results are shown in Figure 3(a) and 3(b), respectively. Ignoring the material loss, the confinement loss is called the partial leakage of the mode power from the core region to the cladding region. The formula for total propagation loss is given by

$$PL = k_0 \, \mathrm{Im}\,(\mathrm{n}_{eff}) \tag{1}$$

The second-order dispersion coefficient (D), which is the most significant one in these evaluations, is given as

$$D(\lambda) = -\frac{\lambda}{c}\frac{d^2}{d\lambda^2}Re(n_{eff}) \tag{2}$$

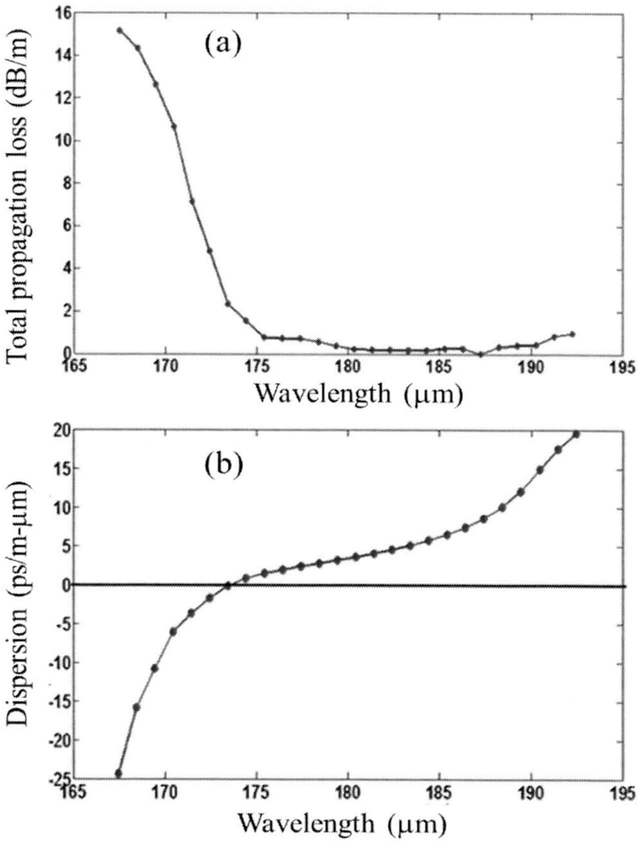

Figure 3. Total propagation loss and dispersion behavior of the PE PBG fiber as described in Figure 2 calculated using the finite element method in the frequency range of 1.56-1.79 THz.

In the above, λ is the wavelength, and $Re(n_{\mathrm{eff}})$ and $Im(n_{\mathrm{eff}})$ are the real and imaginary parts, respectively, of the effective index (n_{eff}).

The FEM includes meshing the cross-section of the PCF structure and assumes a solution by interpolation for the vector wave equation for each mesh. A global mesh system of equations is formed by combining the above-interpolated solutions. Then, under suitable boundary conditions, these equations are solved by using the variational method (Rahman, 2013). By investigating the spectral behavior of the total propagation loss, it is clear that the guiding through such a fiber is appropriate in the wavelength range of 170-195 μm. Dispersion is also low around the center of the transmission spectrum with a zero-dispersion wavelength of $\lambda \sim 174$ μm (frequency of ~1.72 THz), as shown in Figure 3(b). The triangular lattice-based PCF design in Figure 2(c) provides the widest THz bandwidth for guiding among the similar other types of PCF designs considering one air hole per unit cell. Through this work, customization in the PBG-fiber design for the widest possible THz bandwidth with low-loss guidance has been achieved in various polymer host materials (Kumar, Optics and Photonics: Theory and Computational Techniques Workshop, 2018). If a diatomic unit cell is considered in the design of a PBG fiber cross-section, a flat and near-zero dispersion in a broad bandwidth is achievable by choosing different diameters for the two constituents of the basis. Consequently, lower dispersion is also possible (Kumar, IEEE-WRAP, 2019). Similarly, for PBG-fibers with a honeycomb lattice, an even larger bandwidth is possible by increasing the diameter of the central air hole relative to the rest of the air holes in the lattice (Arora, Optics and Photonics: Theory and Computational Techniques Workshop, 2018).

Index Guided Flat and Low Dispersion

Solid core PCFs in Teflon type-2 ($n_2 = 1.44$) host material are discussed here. The cladding region has air holes in a triangular lattice while the near core region (within the first ring) has customized configurations of alternate Teflon type-1 ($n_1 = 1.4$) inclusions and air holes (Kumar, International Conference of Fiber Optics and Photonics India, 2018). Various patterns/configurations are chosen depending on the arrangement of the Teflon type-1 inclusions and the air holes within the first ring around the core. The diameter of these inclusions and air holes in the first ring was varied between 0.54 times the pitch to 0.94 times the pitch for investigating the dispersion characteristics as shown in Figure 4 for four such configurations. The first configuration is without the Teflon-type 1 inclusions in the first ring around the core (Figure 4(a)), the second configuration (Figure 4(b)) contains six Teflon type-1 inclusions symmetrically placed around the core, the third configuration (Figure 4(c)) contains 4, and the fourth configuration (Figure 4(d)) contains 3 of those Teflon type-1 inclusions in the first ring around the core. The corresponding dispersion characteristics are found to be drastically different from one another and for each, the dependence on diameter/pitch ration systematically varies. For all of these, the pitch was set to be 300 μm and the air hole diameter was kept at 282 μm in the cladding region. Depending on the type of configuration as above, these PCFs support the fundamental mode in a broad range from 50 μm to 600 μm. For calculating the dispersion characteristics in Figure 4, the usual beam propagation method was applied, which is described briefly below.

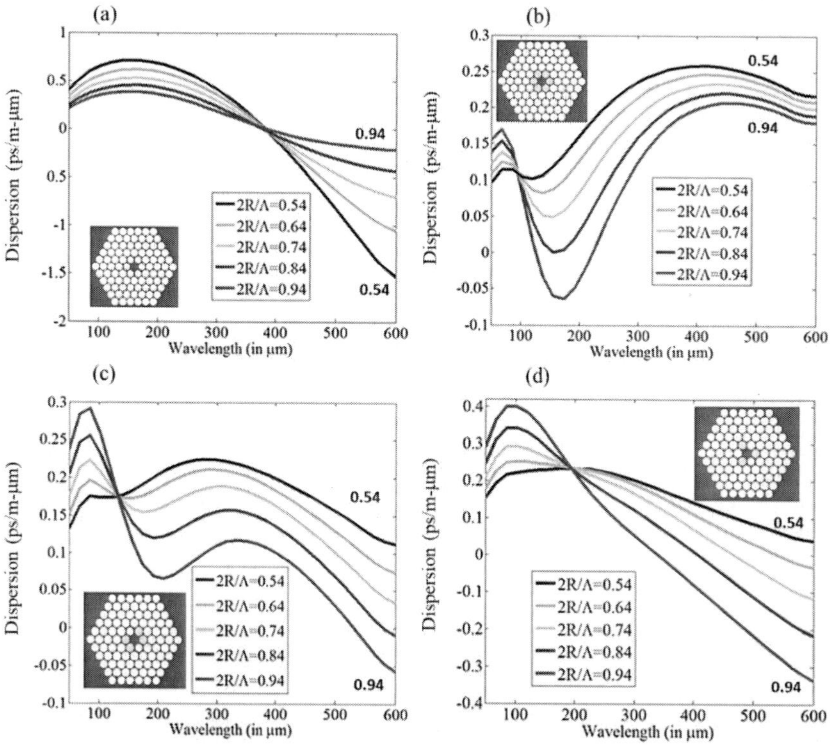

Figure 4. Flexible dispersion characteristics of a Teflon type-2 host (violet region) based a solid core PCF. The white dots in the cladding region represent air columns. The near core region has been configured with Teflon type-1 inclusions (yellow dots) and correspondingly the dispersion characteristics are shown for 4 configurations in (a) to (d). R is the radius of the air and Teflon inclusions, and Λ is the pitch of the pattern in the lattice.

Let $\Psi(x,y,z)$ represent the general solution of the wave equation in 3D space where the wave vector can depend on all three directions. If the whole space is divided into grids of small sizes $(\Delta x, \Delta y, \Delta z)$ then the solution can also be written in the following form,

$$\begin{aligned}\Psi(x,y,\mathrm{z}) &= \Psi(\mathrm{p}\,\Delta x, \mathrm{q}\,\Delta y, r\Delta z)\\ &= \Psi^{r}_{\mathrm{p,q}}\end{aligned} \tag{3}$$

where, integers p, q, and r are the number of steps to be taken to reach *x*, *y*, and *z*, respectively, from the origin. The propagation direction is along the *z*-axis and the field profile is in the *xy*-plane. For demonstration purpose, one can consider a simple 2D Helmholtz equation for field *u* as a function of one transverse (*x*) and another, the longitudinal (*z*) coordinates, given as

$$2in_0k_0\frac{\partial u}{\partial z} = \frac{\partial^2 u}{\partial x^2} + k_0^2(n^2 - n_0^2)u\,. \tag{4}$$

In Eq. (4), n_0 is the refractive index, considered to be an average of the core index and effective index of the clad structure, $n = n(x, z) = n(x)$ is the refractive index profile and k_0 is free space wave vector. Equation 4 can be written as the difference between two wave equations corresponding to forward and backward propagation directions as represented by Eqns. 5a and 5b in the below (Pedrola, 2015), where the variation of the field along the *x*-direction is written in terms of fields at two consecutive *z* points (two consecutive *r* steps),

$$2in_0k_0\frac{u_p^{r+1} - u_p^{r}}{\Delta z} = \frac{u_{p-1}^{r} - 2u_p^{r} + u_{p+1}^{r}}{(\Delta x)^2} + k_0^2[(n_p^{r})^2 - n_0^2]u_p^{r} \tag{5a}$$

$$2in_0k_0\frac{u_p^{r+1} - u_p^{r}}{\Delta z} = \frac{u_{p-1}^{r+1} - 2u_p^{r+1} + u_{p+1}^{r+1}}{(\Delta x)^2} + k_0^2[(n_p^{r+1})^2 - n_0^2]u_p^{r+1} \tag{5b}$$

Due to the continuity of the field in the *z*-direction assumed through the above two equations, a transcendental equation for the *u*-field is obtained as

$$a_p u_{p-1}^{r+1} + b_p u_p^{r+1} + c_p u_{p+1}^{r+1} = d_p \tag{6}$$

Here, a_p, b_p, c_p, and d_p are defined in conjunction with Eqn. 5(a-b). The numerical solution proceeds with a guess modal field function that is close to the actual one (eigen solution) at the input facet of the fiber. Considering all the p-1, p, and p+1 points, a tridiagonal matrix system of n linear transcendental equations is formed, which is then solved for the possible eigen solutions at each point along the z-direction.

The description of the full vector beam propagation method in 3D is quite exhaustive, however, the exact principle as above in the three dimensions is followed (Pedrola, 2015) Commercially available software, RSOFT uses the full vector beam propagation method in 3D to solve the wave propagation through the PCF structures. For the four PCF configurations discussed above, we have obtained the dispersion characteristics as shown in Figures 4(a) to 4(d). During the evaluation of the modal fields, effective index tolerance of 10^{-7} was considered. It can be seen from the results in Figure 4 that by choosing the appropriate configuration, one can engineer the desired zero-dispersion wavelength(s) and flat dispersion near the zero-dispersion line that are helpful in obtaining dispersion compensation, large modal area, and high-power transmission. (Kumar, Springer Proceedings in Physics, vol. 258 (ICOL 2019), 2019).

High Birefringence

Birefringence is the difference between the effective indices of x- and y-polarized modes in a given optical material or fiber. It is an important characteristic of the fibers that can be helpful in various applications such as polarimetry (Xu, 2020). Photonic crystal fibers containing air hole lattices in the clad region can be customized to obtain desired birefringent properties. Below, two specific PCF designs are discussed which, possess the birefringent property at the THz frequencies.

The first design is based on elliptically shaped air holes distributed in a triangular lattice as shown in Figure 5(a). Such a PCF structure though provides high birefringence, but it may be difficult to fabricate them and maintain the same asymmetry in the elliptical air holes throughout the cross-section as well as the entire length of the fibre. Therefore, the second one with circular air holes as shown in Figure 5(b) is a better choice. The corresponding dimensions of the unit cell structures are shown in Figure 5(c). These dimensions are considered such that the sum of the area of the two circular air holes and the area of one elliptical air hole in the triangular lattice are equal (Arora, XLII Optics Society of India, International Symposium of Optics, 2018). This will ensure the same porosity in the two structures, but the latter is more practical and easier to fabricate. In Figure 5(a) and 5(d), the fundamental modal field profiles at 2 THz and the spectral birefringence property through the calculation of effective indices for the x- and y-polarized modes have been presented for the two solid core PCFs by using full vector beam propagation method. It can be seen from Figure 5(c) that the birefringence in both the PCFs is nearly comparable, though the structure with circular air-holes can be further optimized to achieve higher birefringence.

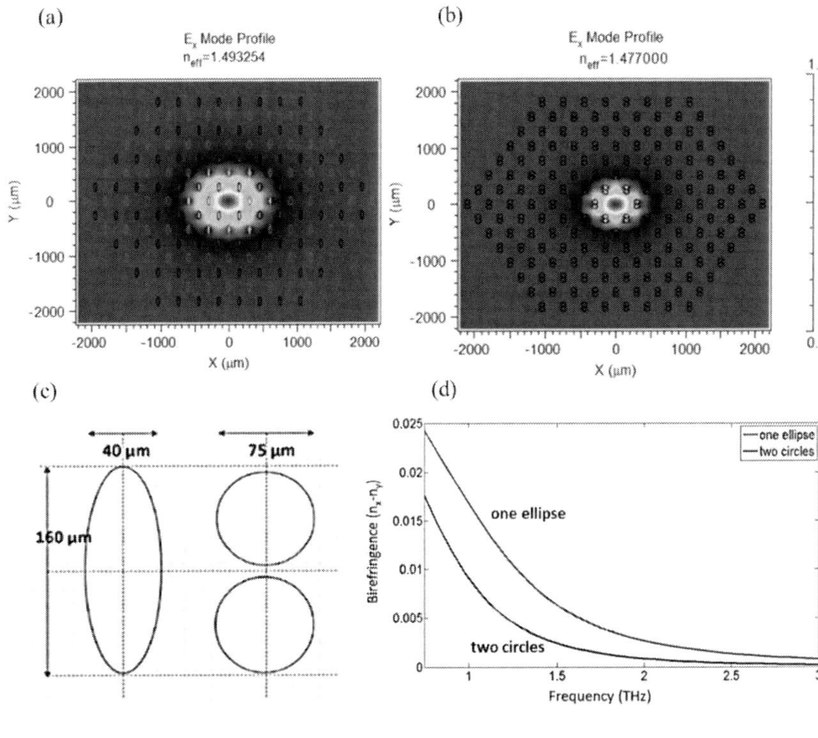

Figure 5. Effect on the birefringence and modal profile of a solid core birefringent PE PCF due to the variation in the basis-type of the triangular lattice configuration in the cladding. (a-b) The lattice configurations and the modal profiles, (c) basis type, either a single elliptical or two circular air columns used in the lattice configuration, and (d) birefringence for the two basis types.

For the above results, the pitch was kept at 200 μm for both the structures, the size of the elliptical hole in the first was 160 μm/40 μm and that of the circular hole in the second was 75 μm. The modal field appears more confined in the core in the second structure with circular holes than in the first one with elliptical holes. Furthermore, as expected, the birefringence is the highest at near 0.5 THz and vanishes at high frequencies beyond ~2.5 THz.

Polarization mode dispersion (PMD) is an important property that quantifies the difference between the dispersion of the PCF for its *x*- and *y*-polarized modes. From the spectral variation of the effective indices, the PMD for both the above PCFs was found to be low. For certain applications, low PMD may be a good choice in maintaining high birefringence as well as proper guidance throughout the length of the fiber. PCF designs can be adopted in a way that positive or negative PMD can be obtained such that they can be utilized in PMD compensation by combining two or more such fibers with opposite PMD (Menyuk, 2006). Moreover, by adjusting the solid core structure, it is possible to enhance the birefringence and the PMD properties of the PCFs (Arora, International Conference of Fiber Optics and Photonics India, 2018).

THz PCFs as Polarization Splitters

Polarization control and detection at THz frequencies by suitable optical elements including fibers and other devices is a developing field. The currently available THz polarizers and other polarization devices are not cost-effective and, also work mostly for a narrow range. It is desirous to design PCF-based polarization-sensitive devices which provide better control and manipulation of the THz wave polarization in a broad frequency range so that polarization-sensitive measurements and applications can be carried out with a specific state of polarization (Koch, 2014; Hidayat, 2008; Burrow, 2017; Morris, 2012; Jena, 2020; Guo, 2020). Fiber-based polarization splitters are an upcoming field for applications, where new fiber designs to optimize specific outcomes are being put to work. Using cascaded multimode interference fiber structures, the unpolarized input beam is divided into TE and TM polarizations as it travels through the fiber length. To achieve a certain polarization extinction ratio, a polarization splitting length is defined for such fibers which, can work either in a narrow or a broad frequency range depending on the design and the material used in the fiber (Li, 2014; Sesay, 2013; Hibbard, 2019; Park, 2010; Zhong, 2015).

Dual Core Configuration

Photonic crystal fibers or the microstructured fibers with dual-core configurations have been the subject of an intense study recently due to their usefulness in engineering desired dispersion and polarization properties at THz frequencies. By optimizing the structure of the fiber, THz wave coupling, polarization splitting, dispersion control, etc. can be achieved within a short length, i.e., the device length. For low-loss at THz frequencies, all the above fibers are invariably made of polymers. For example, a suspended dual elliptical-core microstructured fiber in Topas host material can be used as a coupler and it provides polarization splitting with different splitting lengths for two polarization states (Chen, 2016). Such a structure with an elliptical core has certain limitations. Foremost, maintaining the same ellipticity along the length of the fiber is a great challenge from the fabrication point of view. Shorter splitting lengths and better extinction ratio in the polarization splitting can be achieved by using other simpler designs, one of which is described below in the next section.

Microstructured Dual Core with Air Holes

Polyethylene as a host material in THz PCFs is suitable because it is transparent in a large THz frequency range. PE PCF-based THz polarization splitters have been reported in the literature, however, the splitting length is quite large and the operational frequency range is also narrow (Li, 2013). PE dual-core microstructured fibers with a special configuration of air hole arrangement in the two cores are designed and the corresponding modal field and other characteristics have been discussed here. The specific goals of this study were to shorten the splitting length and enhance the extinction ratio and the operational frequency bandwidth. In the first structure (Str-1) as shown in Figure 6, the two cores have been configured for a gradience in the pitch with which the air-holes in the lattice are distributed. This type of design helps in engineering spatial variation in the refractive index to achieve specific profiles of the effective indices. As shown in Figure 6, the clad has a hexagonal ring lattice of air holes with pitch, Λ = 400 μm, and diameter, d = 340 μm. The pitches inside the cores are Λ_1 = 60 μm, Λ_2 = 70 μm and Λ_3 = 80 μm, respectively. The diameters of air holes in the core are d_1 = 40 μm and d_2 = 35 μm, respectively. This dual-core PE PCF has been optimized for THz polarization splitting in the frequency range of 0.4-1 THz. Designing gradient pitches in the arrangement of air holes provides more flexibility in effective indices vs. frequency curve which is discussed below in detail.

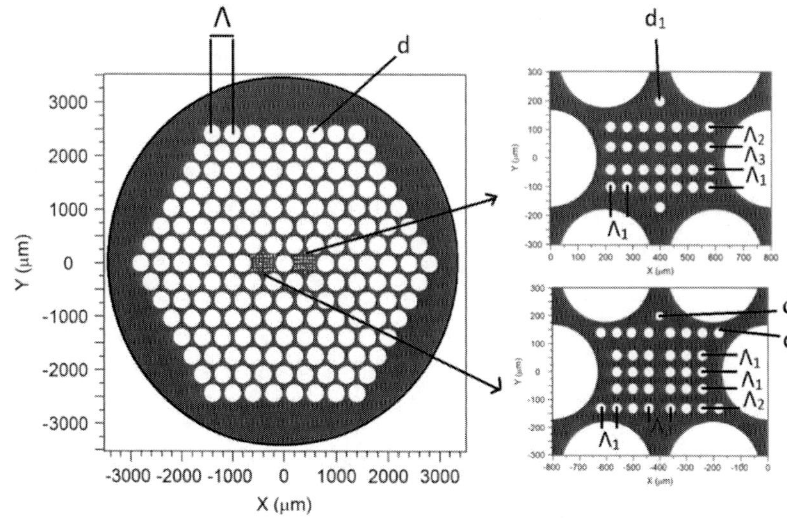

Figure 6. Schematic of the transverse cross-section of a PE-based dual-core PCF (Str-1) along with the magnified views of the two cores having two different air hole configurations. The red color represents the PE host material, and the white dots represent air columns in the fiber. Various pitches (Λ) and diameters (d) are marked. $\Lambda = 400$ μm, $d = 340$μm in the cladding region; $\Lambda_1 = 60$ μm, $\Lambda_2 = 70$ μm, $\Lambda_3 = 8$ μm, $d_1 = 40$ μm, $d_2 = 35$ μm in the dual-core region.

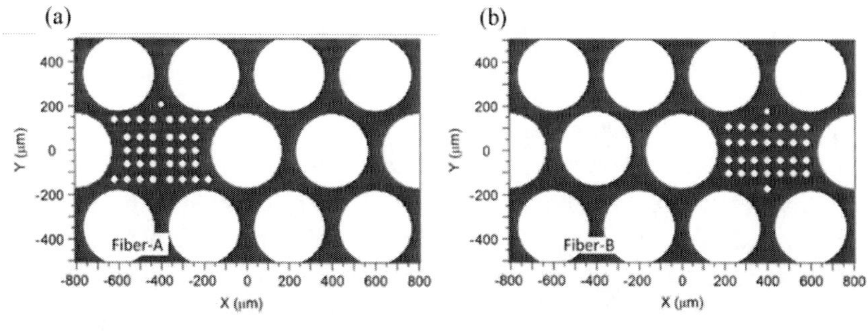

Figure 7. Magnified views of the two cores considered in the dual-core PE PCF. (a) Dual-core PE PCF: Fiber-A consisting of air-column configuration in the left core, while the right core is simply an air column of diameter same as in the outer region, (b) dual-core PE PCF: Fibre-B consisting of air-column configuration in the right core, while the left core is simply an air column of diameter same as in the outer region. The diameters and pitches of the air columns in the Fiber-A and Fiber-B are already shown in Figure 6.

For a complete understanding of the underlying polarization splitting mechanism, three configurations for this scheme are considered. (1) Fiber-A consists of a special air-column configuration in the left core as shown in Figure 7(a), while the right core contains a simply an air column of diameter same as that in the outer region. (2) Fibre-B consists of another special air-column configuration in the right core as shown in Figure 7(b), while the left core is simply composed of an air column of diameter same as in the outer region. (3) Fiber-AB is one which has left core configuration as that of fiber-A and right core configuration as that of fiber-B as described earlier and shown in Figure 6.

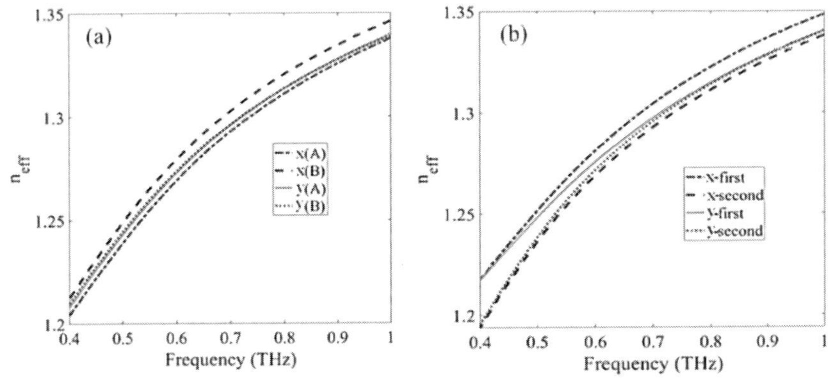

Figure 8. (a) Effective indices of the x- and y-polarized modes of the PE-based dual-core Fiber-A and Fiber-B. (b) Effective indices of the first two modes for the x- and y-polarized modal fields in the PE-based dual-core PCF, Fiber-AB as described in Figure 6 consisting of two cores with air-hole configurations shown in Figure 7.

By applying full vector beam propagation method, various characteristics of the above three fibers, viz, Fiber-A, Fiber-B and Fiber-AB have been determined in the frequency range of 0.4-1 THz. The effective index profiles are presented in Figure 8. The effective indices for one of the polarizations, i.e., *x*-polarized modes in the individual two cores-based fibers, i.e., Fiber-A and Fiber-B, differ massively as shown in Figure 8(a). On the other hand, the effective indices for other polarization, i.e., *y*-polarized modes, are nearly degenerate around a crossing frequency of ~0.85 THz. It can be seen that the effective indices of the *y*-polarized modes in individual cores are close to each other with a crossing frequency equal to 0.85 THz. For the effective index profiles for x- and y-polarizations in Figure 8(b): *x*-first, *x*-second, *y*-first, and *y*-second correspond to the first and second modes of the Fiber-AB for its x-

and y-polarizations, respectively. This structure has been optimized to achieve an appropriate effective index profile for a short splitting length.

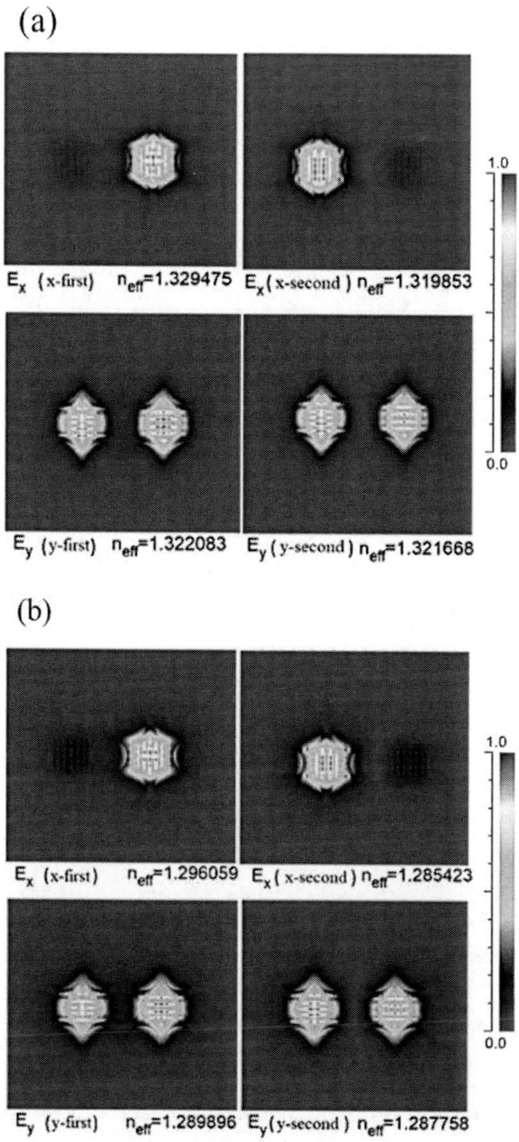

Figure 9. *x*- and *y*-polarized modal-field profiles at (a) 0.85 THz and (b) 0.66 THz for the first and second modes of the PE-based dual-core PCF described in Figure 6 calculated using the full-vector beam propagation method.

Figure 9 presents the modal field profiles calculated using the full vector beam propagation method for the fiber design, Fiber-AB, as described in Figure 6, where, both the first and second modes for the x- and y-polarizations are shown at two representative frequencies: 0.85 THz in Figure 9(a) and 0.66 THz in Figure 9(b). It can be seen that the modal fields for the y-polarization are present in both the cores. This coupling has occurred due to the small difference between the effective indices for the y-polarization in Fiber-A and Fiber-B. On the other hand, for the x-polarization case, the fields remain within the individual cores. Also, with the increase in the wavelength of the THz wave, the confinement in the core region gets reduced.

The difference between the y-first and y-second effective indices is used in the evaluation of the splitting length/device length, which is given by the relation,

$$L_c = \frac{\lambda}{2(n_{eff-y1} - n_{eff-y2})} \tag{7}$$

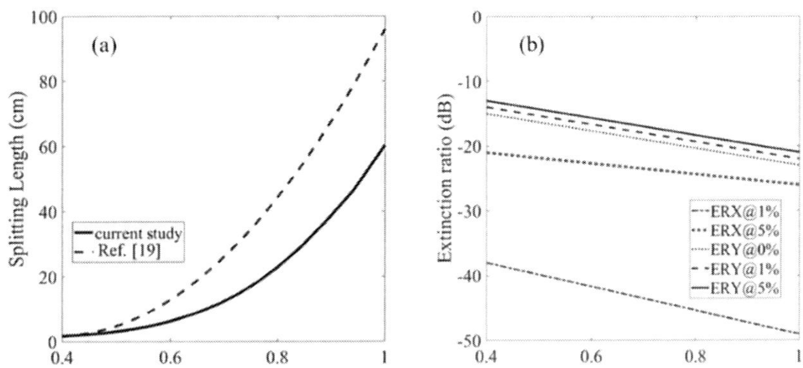

Figure 10. (a) Frequency-dependent polarization splitting length or the device length of a polarization splitter based on the dual-core PE PCF described in Figure 6. (b) Frequency dependence of the polarization extinction ratio in x-polarization (ERX) and y-polarization (ERY) for certain % deviation in the device length.

The effective indices of y-polarized modes in Fiber-AB differ at low frequencies but are quite close at high frequencies. Consequently, the polarization splitting length becomes larger with the increasing frequency. The splitting length increases from ~1.96 cm at 0.4 THz to~ 60 cm at 1 THz. In

Figure 10(a), the splitting length of Fiber-AB has been compared with another design from the literature (Li, 2013). Clearly, Fiber-AB design is more appropriate for constructing compact polarization splitting devices (Kumar, Applied Optics, 2020). The gradience in the air holes configurations in the two cores of Fiber-AB is responsible for the tunability of the effective indices and hence the compact splitting length.

Extinction ratio signifies the efficiency of polarization splitting at the splitting length. The extinction ratios ERX & ERY for the *x*- and *y*-polarized modes, respectively can be defined as follows.

$$ERY = 10\log_{10}\frac{P_{out}(X(B))}{P_{out}(Y(B))} \tag{8a}$$

$$ERX = 10\log_{10}\frac{P_{out}(Y(A))}{P_{out}(X(A))} \tag{8b}$$

Here, $P_{out}(X(A))$, for a definition, means the output power of *x*-polarization in core-A of the Fiber-AB. The power transferred from core-A to core-B for y-polarization is given by the relation,

$$P = P_0 Sin^2\left(\frac{\pi}{2}\frac{L}{L_c}\right) \tag{9}$$

The extinction ratios for the PE-based dual-core PCF are presented in Figure 10(b) for certain % deviations of the device lengths with respect to the splitting length. The values are significantly large as desired in real experiments. The extinction ratios for *y*-polarization vary in the range of ~15-23 dB in the entire frequency range under consideration, i.e., 0.4THz to 1 THz. Similarly, in the given frequency range, the extinction ratios for *x*-polarization vary only in the range of ~21-25 dB. A slight decrease in the values of the extinction ratios with the frequency is due to the increase in the *x*-mode confinement.

Microstructured Dual Core Composite

Selective placement of Teflon inclusions in an optimized configuration of air holes in the core region of a dual-core PE PCF as discussed in the previous section can lead to more desirable characteristics for THz polarization splitting (Kumar, Results in Optics, 2021). Figure 11(a) describes the structure (Str-2) of such a dual-core polyethylene-based photonic crystal index-guided fiber with the details of the two cores, given in Figures 11(b) for core-A and 11(c) for core-B. There is a difference between the air-hole configuration and the Teflon inclusions in it within the two core structures shown in Figures 11(b) and 11(c). Like before, the fiber consisting of only core-A and an air column simply at the place of core-B is named as Fiber-A. Similarly, the fiber with only core-B is named as Fiber-B and the one with both the cores in place is named as Fiber-AB. The pitch, Λ in the clad is 400 µm, and the air hole diameter, d, is 340µm. Various pitches, $\Lambda_1 = 60$ µm, $\Lambda_2 = 70$µm, $\Lambda_3 = 80$µm and diameter of air holes, $d_c = 40$µm, have been considered in the core regions. These structural parameters in the cores and their configuration are taken such that effective indices of two polarized modes become more suitable for better performance in polarization splitting. By inclusion of the Teflon columns in the air-hole configuration in the two cores, it helps to achieve almost zero difference between the n_{eff} values of y-polarized modes in the individual cores (Fiber-A and Fiber-B) and hence short polarization splitting device length.

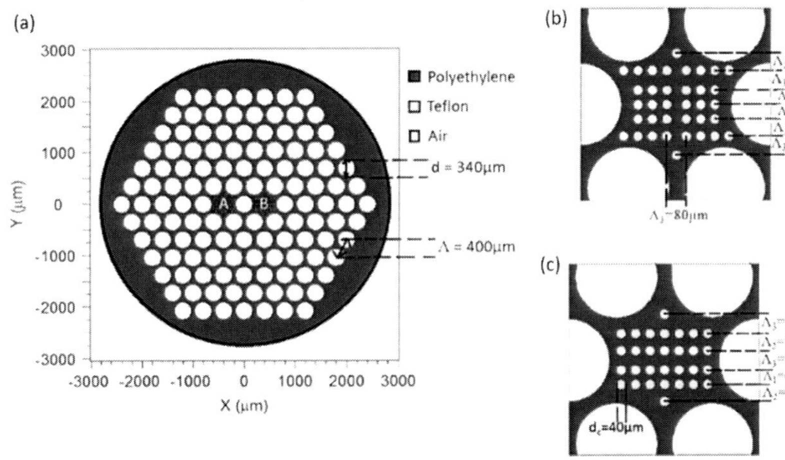

Figure 11. (a) Transverse cross-sectional view of the PE-based dual-core microstructured fiber: Fiber-AB (Str-2) containing air and Teflon inclusions; the two cores have been identified as A and B. Magnified views of (b) the core A, and (c) the core B regions of the dual core fiber. Different pitch values (Λ) and diameters (d) of the Teflon and air columns are indicated. All remaining pitches in the x-direction are the same, i.e., $\Lambda_1 = 60$µm.

Using the full vector beam propagation method, calculation of the effective indices in individual cores-based fibers, i.e., Fiber-A and Fiber-B and the dual core fiber, Fiber-AB have been carried out in the frequency range of ~0.4-1 THz for the x- and y-polarizations. The corresponding results are presented in Figure 12(a). The interesting point to note from these results would be to compare the differences in the effective index profiles of the two structures of the Fiber-AB, i.e., Str-1 in Figure 6 and Str-2 in Figure 11. The first thing to note from the results in Figure 12 is that the effective indices for the x-polarization are quite different from each other for the three fibers designs, i.e., Fiber-A, Fiber-B and Fiber-AB. On the other hand, they are quite close to each other for the y-polarization. It can be seen from Figure 12(a) that the effective index profiles for y-polarization for the Fiber-A and Fiber-B intersect at a frequency of ~0.65 THz. This frequency of intersection was at ~0.85 THz for Fiber-A and Fiber-B of Str-1 in Figure 8(a). Same thing is found for the dual-core fiber, Fiber-AB if results are compared from Figure 12(b) for Str-2 and Figure 8(b) for Str-1. In other words, by having Teflon inclusions in the air-hole configuration of the dual-core PE PCF, Fiber-AB with Str-2, the frequency of minimum difference between the effective indices for y-polarization shifts to a lower value than what was obtained for Fiber-AB with Str-1. Consequently, short splitting length for Fiber-AB with Str-2 could be achieved at higher frequencies also as discussed later in Figure 14(a).

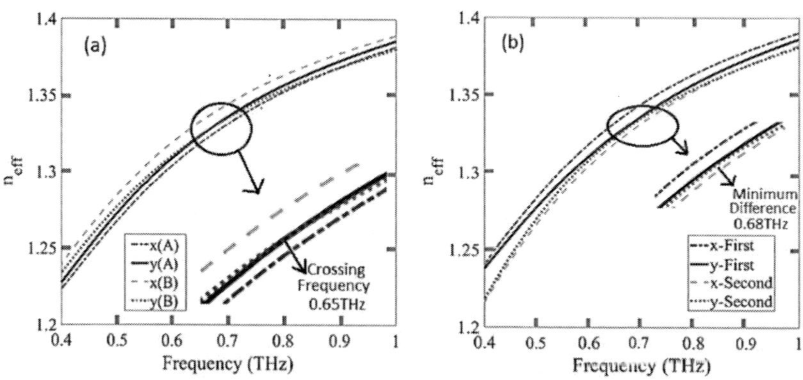

Figure 12. Effective indices, n_{eff} for the x- and y-polarizations of (a) Fiber-A and Fiber-B, and (b) the dual-core fiber, Fiber-AB as described in Figure 11 (Str-2). Inset in (a) is the magnified view around the crossing frequency at ~0.65 THz. Inset in (b) is the magnified view to better see the difference among the curves in the frequency.

The modal field profiles as calculated by using full vector beam propagation method for x-and y-polarizations in the dual core Fiber-AB with Str-2 are presented in Figure 13. These are calculated for THz wave frequencies of 0.6, 0.7, and 0.8 THz with the corresponding results shown in Figures 13(a), 13(b), and 13(c), respectively, where the values of the effective indices are also included. These results suggest that there is least coupling of power between the two cores for the x- and y-polarizations. From the modal field size it is clear that the confinement is more for higher frequency of the input THz field.

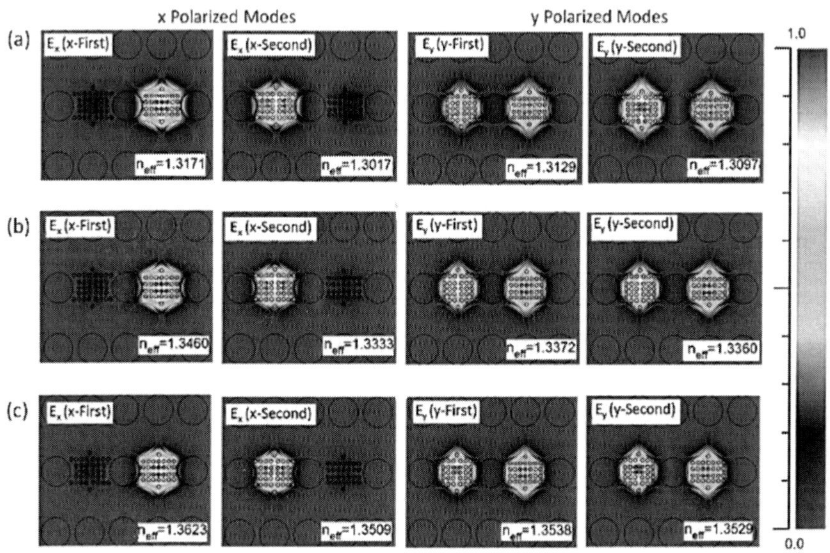

Figure 13. First and second modal electric field distributions in the dual core microstructured fiber, Fiber-AB (as described in Figure 11) for the *x*- and *y*-polarizations calculated at a frequency of (a) 0.6 THz, (b) 0.7 THz, and (c) 0.8 THz. The color bar represents the variation in the electric field strength.

The polarization splitting length of the Fiber-AB with Str-2 as shown in Figure 11 is shown in Figure 14(a) for the frequency varying from 0.4 THz to 1 THz. For a comparison, results for Str-1 as presented in the previous section and that for another similar structure from literature, are also shown in the inset of Figure 14(a). Clearly, short splitting length is obtained for this device even at higher frequencies, in the range of ~0.7-1 THz, unlike the case of Fiber-AB with Str-1 for which the splitting length increases with the increasing frequency. It has already been seen in Figure 12(b) that the difference between the effective indices goes minimum at frequency of ~0.68 THz, corresponding to which, there is a maximum that can be seen in the behaviour of the splitting length. Beyond 0.68 THz, the splitting length decreases continuously to reach a minimum value close to one at the low frequencies. The splitting length is ~2.1 cm at 0.4 THz, goes to a maximum of 20.1 cm at ~0.68 THz and then reduces again to ~2.1 cm at 1 THz. Therefore, such a polarization splitter of a fixed length can be used effectively for a broad frequency range. In Figure 14(a), we have also compared the outcome with another case of the Fiber-AB in which, there are Teflon inclusions in core A only in the two layers at the center that is to make a slit-like pattern (see Figure 11(b)). By doing so, the splitting length becomes larger at high frequencies.

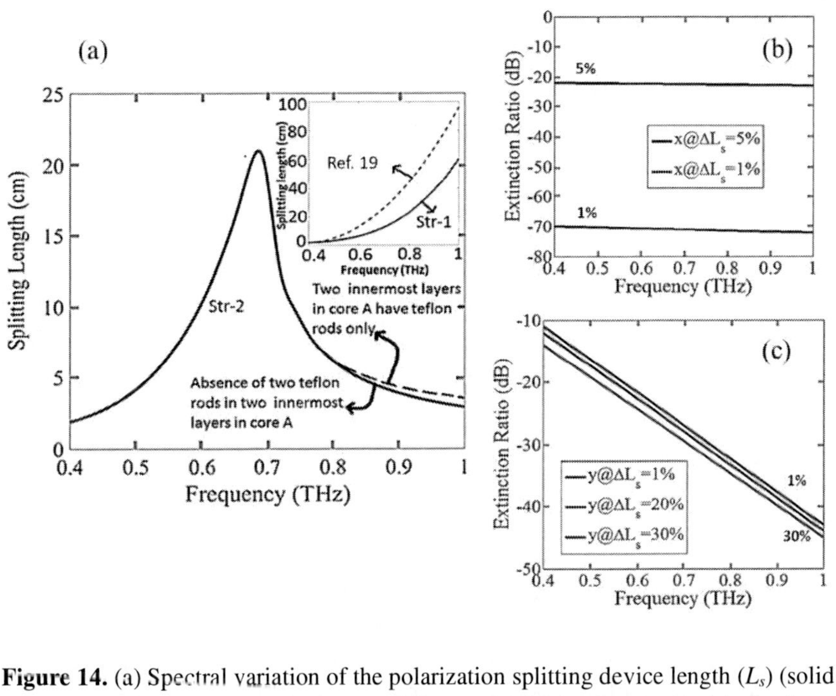

Figure 14. (a) Spectral variation of the polarization splitting device length (L_s) (solid curve): the superimposed dashed curve (bifurcated at higher frequencies) is for the structure when two innermost air holes in the last row of core-A (see Figure 11) are replaced with Teflon rods. Inset: A comparison between the current results for Str-2 of the PCF and those for Str-1 taken from Refs. (Li, 2017) and (Kumar, 2020). Spectral variation of the extinction ratio for (b) *x*-polarization and (c) *y*-polarization at different percentage deviations (ΔL_s) in the device length.

The polarization extinction ratios for the *x*- and *y*- polarizations are presented in Figures 14(b) and 14(c), respectively, for different % deviations in device length from actual splitting lengths. It is apparent that the extinction ratio for the *x*-polarization is highly sensitive to deviation in the device length. The reason behind this is the fact that the power in core-A varies sinusoidally with the device length due to coupling to the other core for the *y*-polarization. From Figure 14(c), it can also be seen that the extinction ratios for the *y*-polarization are sensitive to frequency instead of the deviation in the device length. This arises because for x-polarization, there is low power coupling to the core-B and hence low output at the end of the device.

Generation of THz Radiation by Four-Wave Mixing in PCFs

Nonlinear fiber optics-based devices for high-power THz wave generation is an emerging field (Suizu, 2007; Zhou 2011; Qasymeh, 2012; Barh, 2015). In the limited reports available in the literature, waveguides of silicon, silica, and Teflon have been taken for the generation and guidance of THz waves (Qasymeh, 2012; Barh, 2015; Brothers, 1994; Demas, 2017). Difference frequency generation (DFG) and four-wave mixing (FWM) phenomena using infrared lasers in the conventional silica fibers (Suizu, 2007) and silicon-based PCIGF (Demas, 2017) have produced THz waves in the low-frequency region. Silicon step-index conventional fiber could produce a THz generation efficiency of only 10^{-5} due to strict noncolinear phase matching for the DFG process at low frequencies (Suizu, 2007). With co-linear phase matching in silicon step-index fiber, the THz generation efficiency showed an improved value of $\sim 10^{-3}$ (Zhou 2011; Qasymeh, 2012).

The use of polymers can make it possible to achieve a much larger difference in the outcomes. Using four-wave mixing process for the THz wave generation in a Teflon PCF, an efficiency of ~3% at 3.14 THz was reported recently (Barh, 2015), however, it requires a long fiber length. Microstructured inclusions-based step-index core and Teflon as the host material and by using its variants for the inclusions, different designs of the PCF have been studied, which provide good control over the dispersion property of the PCFs to tune the frequency and generation efficiency of the output THz radiation (Jauregui, 2012; Kumar, Optics communication, 2020; Agarawal, 2007; Boyd, 2008). A scheme has been discussed in the following section where efficient THz generation can be achieved by FWM in a Teflon-based PCF within a much shorter length than reported before.

Nonlinear Wave Mixing in a PCF

Four-wave mixing (Agarawal, 2007) is the third-order nonlinear optical process employing two pump waves and one idler wave that propagate through a medium of high third-order nonlinear optical susceptibility. The three input waves mix and produce the fourth wave in a specific direction depending on the phase matching condition. The energy and momentum of the photons involved are satisfied necessarily, in all such processes. In a PCF, the interaction among the four waves is much easier and interaction lengths can be much larger, which basically helps to improve the four-wave mixing process and hence the intensity of the generated radiation. Below, we consider two cases of FWM in different designs of the PCFs which can produce highly efficient THz output in a narrow frequency range and in a broad frequency range.

Narrow-Band Nonlinear PCF THz Source

A microstructured fiber using a Teflon variant (n_c = 1.425) as the core background material and other Teflon variant (n_{cl} = 1.4) as the uniform-clad material (Kumar, Optics communication, 2020) has been designed. The cross-section of the microstructured fiber is shown in Figure 15(a). The radius of the core is R_{core} = 150μm and that of the clad is R_{clad} = 185 μm. The core has embedded 1 μm diameter (d) micro-inclusions of Teflon-3 (n_{tf} = 1.44) and TiO_2-doped polyethylene (n_{pe} = 2.1) in alternate hexagonal rings while the separation between two consecutive hexagonal rings of the micro-inclusions is Λ = 30 μm. Such a configuration of the micro-inclusions is equivalent to the hexagonal structure of diatomic basis.

Figure 15(b) shows a large and nearly flat modal field of the infrared pump at a wavelength of λ_p = 10.56 μm, which ensures large overlap with a seed idler wave at the input to result into good overlap with the generated THz modal field and hence high output power generation efficiency in the FEM process. By considering the above core configuration, efficient THz wave output has been produced at a frequency of 9.33 THz, which is same as the neurotransmitter spectroscopic resonance frequency and hence has significant implications. The chosen materials in the fiber design possess high nonlinearity and hence high THz output power is produced via four-wave mixing process. Also, the design in the core ensures good overlap among the interacting fields and proper guidance of all of them over long length of the fiber.

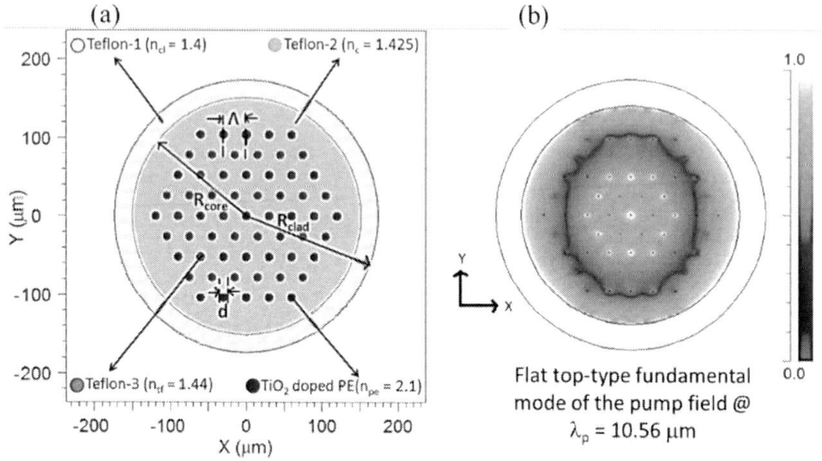

Figure 15. (a) Cross-sectional schematic of the diatomic hexagonal rings based PCF, where Teflon-2 based core consists of alternate hexagonal rings of Teflon-3 and TiO_2 doped PE columns and clad is made of Teflon-1. The refractive indices of various regions have been indicated in different colors. $n_{cl} = 1.40$ is for Teflon-1, $n_{co} = 1.425$ is for Teflon-2, $n_{tf} = 1.44$ is for Teflon-3 and $n_{pe} = 2.1$ is for TiO_2 doped polyethylene. (b) Fundamental mode for infrared pump wave at a wavelength (frequency) of $\lambda_p = 10.56$ μm (28.40 THz) in the microstructured fiber.

Infrared pump at wavelength λ_p (frequency f_p) = 10.56 micron (28.40 THz) can be taken from a CO_2 laser and an input seed idler wave at variable wavelength λ_i (frequency f_i) can be taken from a CO laser (~47.6 THz). The effective pump modal field area, given by Eqn. 10 below is calculated to be $A_{eff} \sim 35500$ μm^2 using the full vector beam propagation method.

$$A_{eff} = \frac{\left(\iint_{-\infty}^{\infty} \left|F(x,y)\right|^2 dxdy\right)^2}{\iint_{-\infty}^{\infty} \left|F(x,y)\right|^4 dxdy} \tag{10}$$

The phase-matching condition for efficient FWM requires the sum of all the linear and nonlinear phase terms to be nearly zero, i.e., $\Delta k = \Delta k_l + \Delta k_{nl} = 0$, where k is the wave vector, Δk_l is linear phase mismatch because of the dispersion in the fiber, and Δk_{nl} is the nonlinear contribution due to self-phase modulation by strong pump field. They are given below.

$$\Delta k_l = k_i + k_s - 2k_p \tag{11a}$$

$$\Delta k_{nl} = 2\Upsilon P_p \tag{11b}$$

$$\Upsilon = \frac{\omega_p n_2}{cA_{eff}} \tag{11c}$$

Here, k_i is the idler wave vector; k_s is the signal THz wave vector, k_p is the pump wave vector, n_2 is the nonlinear refraction coefficient at the pump-frequency ω_p, c is the speed of light in vacuum, and P_p is the pump power. The nonlinear refraction coefficient for the Teflon fiber has been taken to be $n_2 = 8.3\times10^{-18}$ m^2/W (Boyd, 2008). For the calculation of Δk_l, the propagation constant k can be expanded around the pump frequency, ω_p, using the Taylor series expansion method as illustrated in the following equations.

$$\Delta k_l = (k_i - k_p) + (k_s - k_p) \tag{12a}$$

$$k_i - k_p = \frac{1}{1!}\beta_1(\omega_i - \omega_p) + \frac{1}{2!}\beta_2(\omega_i - \omega_p)^2 + \\ + \frac{1}{3!}\beta_3(\omega_i - \omega_p)^3 + \frac{1}{4!}\beta_4(\omega_i - \omega_p)^4 + \ldots.. \tag{12b}$$

$$k_s - k_p = \frac{1}{1!}\beta_1(\omega_s - \omega_p) + \frac{1}{2!}\beta_2(\omega_s - \omega_p)^2 + \\ + \frac{1}{3!}\beta_3(\omega_i - \omega_p)^3 + \frac{1}{4!}\beta_4(\omega_i - \omega_p)^4 + \ldots.. \tag{12c}$$

Here, $\beta_i = \frac{d^i}{d\omega^i}(k_p n_{eff})$ is the ith order dispersion coefficients. Considering $\omega_i - \omega_p = \omega_p - \omega_s$ in the above, the equations are solved to obtain,

$$\Delta k_l = \beta_2(\omega_i - \omega_p)^2 + \frac{\beta_4}{12}(\omega_i - \omega_p)^4 + ... \tag{13}$$

All dispersion effects higher than 4th order have been neglected. Terms contain odd orders of the dispersion have got cancelled out naturally due to opposite signs of the frequency differences. Under perfect phase matching, $\Delta k = \Delta k_l + \Delta k_{nl} = 0$, the following relation is obtained.

$$\beta_2(\omega_p)(\omega_i - \omega_p)^2 + \frac{\beta_4(\omega_p)}{12}(\omega_i - \omega_p)^4 + 2\Upsilon P_p = 0 \tag{14a}$$

or $$\beta_2(\omega_p)(\omega_p - \omega_s)^2 + \frac{\beta_4(\omega_p)}{12}(\omega_p - \omega_s)^4 + 2\Upsilon P_p = 0 \tag{14b}$$

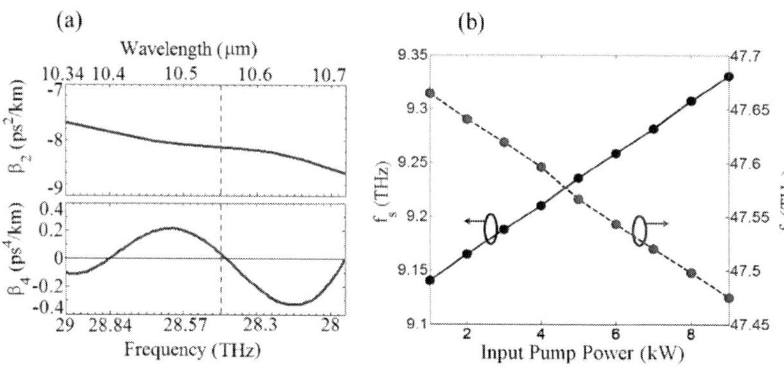

Figure 16. (a) Spectral dependence of dispersion coefficients, β_2 and β_4 around the pump wavelength for the proposed diatomic hexagonal rings-based PCF shown in Figure 15(a). (b) Tuning of THz signal and idler frequencies by variable pump power. The vertical dashed line in (a) represents the pump wavelength of $\lambda_p = 10.56$ μm (frequency, $f_p = 28.40$ THz) that has been considered in the degenerated four-wave mixing process.

The second and fourth order dispersion coefficients around the pump wavelength λ_p (frequency $f_p = \omega_p/2\pi$) = 10.56 micron (28.40 THz) were calculated, and they are presented in Figure 16(a). These results when substituted into Eq. 14 provide the results for the allowed frequencies of the idler and signal (THz) waves for which the phase matching condition is satisfied, and then put in the above Eqns. 14, the formation of the phase-matching equation in terms of idler or signal frequency provided idler or THz signal frequency. At the pump power of 2 kW, the phase-matched idler and signal frequencies are 47.64 THz and 9.16 THz, respectively. In fact, these are tunable by adjusting the pump power as shown in Figure 16(b), where we have varied the pump power from 1 kW to 9 kW and the in accordance with that, the signal and idler frequencies change from 9.14 THz to 9.33 THz and from 47.68 THz to 47.48 THz, respectively.

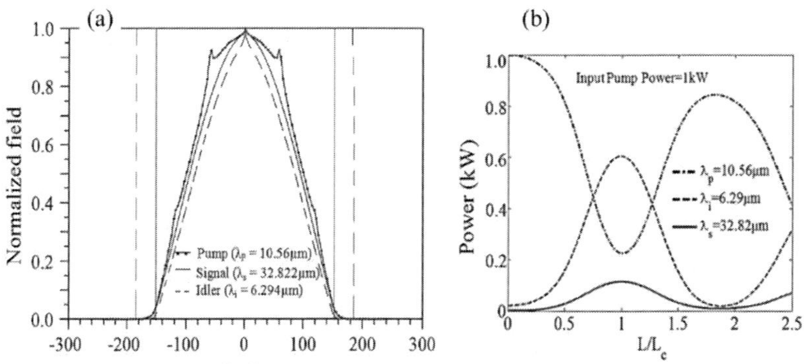

Figure 17. (a) Normalized fundamental modal field variations in the radial (r) direction for the pump field amplitude A_p, idler field amplitude, A_i, and THz signal field amplitude A_s used in the degenerated four-wave mixing in the PCF as shown in Figure 15(a). The corresponding wavelengths are indicated. (b) Pump, idler, and THz signal power dependence on the fiber length, L. Pump power of 1 kW and idle input power of 20 W was. The coherence length, L_c, and the optimum THz signal frequency obtained from the phase-matching condition are 37.8 m and 9.14 THz. Here nonlinear absorption has been ignored, i.e., $\alpha_{2x} = 0$.

The dielectric strength of Teflon is quite large, of ~24 MV/m (R J Chase company, PTFE); it is safe to have the pump power as high as 15 kW at the effective mode area $A_{eff} = 36{,}000\ \mu m^2$. Using good pump power of 15 kW, the corresponding pump electric field is ~14.84 MV/m which is much lower than the dielectric breakdown of Teflon. Therefore, while analyzing the pump-power dependent tunability in the signal and idler frequencies and the THz generation efficiency in the FWM, the pump power is varied upto the maximum of 15 kW only. We point out that the Raman scattering at the chosen pump frequency is insignificant in Teflon (Mihály, 2006); therefore, large input pump power can be taken, and its efficacy leads to shorter optimized fiber length for maximum signal output. Moreover, the Brillouin scattering is insignificant as the corresponding frequency shift of 2.39 GHz is very small (R J Chase company, PTFE) as compared to pump frequency.

The modal profiles of the pump, idler, and signal THz waves have been calculated using the full vector beam propagation method and they are presented in Figure 17(a). A good overlap among the three fundamental modal profiles can be seen; the chosen wavelengths of the three fields for this demonstration are λ_p = 10.56 μm, λ_s = 32.822 μm and λ_i = 6.294 μm, respectively, as shown in Figure 17(a). The factor *f*, representing the modal field overlapping as described in the following paragraphs, is found out to be large enough to have an efficient THz output generation from the PCF.

The wave equations for the three fields viz, the pump, the idler, and the signal, can be derived from Maxwell's equations. Nonlinear equations are then written down for each one of them by taking into account properly the nonlinearly amplified or attenuated amplitudes, the self and cross-phase modulations, and four-wave mixing terms. The interacting four waves, i.e., two pump waves, one noisy idler wave, and one output THz wave, all propagating along the fiber length obey the coupled equations as follows,

$$\frac{dA_p}{dz} = -\frac{\alpha_{1p} A_p}{2} + \frac{i n_2 \omega_p}{c} \left\{ \left(f_{pp} |A_p|^2 + 2 \sum_{k=i,s} f_{pk} |A_k|^2 \right) A_p + \right.$$

$$\left. + 2 f_{ppis} A_p^{*} A_i A_s e^{j \Delta k_l l} \right\} - \alpha_{2p} A_p A_p^{*} A_p \tag{15a}$$

$$\frac{dA_i}{dz} = -\frac{\alpha_{1i} A_i}{2} + \frac{i n_2 \omega_i}{c} \left\{ \left(f_{ii} |A_i|^2 + 2 \sum_{k=p,s} f_{ik} |A_k|^2 \right) A_i + \right.$$

$$\left. + f_{ispp} A_s^{*} A_p A_p e^{-j \Delta k_l l} \right\} - \alpha_{2i} A_i A_i^{*} A_i \tag{15b}$$

$$\frac{dA_s}{dz} = -\frac{\alpha_{1s} A_T}{2} + \frac{i n_2 \omega_s}{c} \left\{ \left(f_{ss} |A_s|^2 + 2 \sum_{k-p,i} f_{sk} |A_k|^2 \right) A_s + \right.$$

$$\left. + f_{sipp} A_i^* A_p A_p e^{-j\Delta k_i l} \right\} - \alpha_{2s} A_s A_s^* A_s \tag{15c}$$

$$f_{ab} = \frac{\int_{-\infty}^{\infty} |F_a(r)|^2 |F_b(r)|^2 r dr}{2\pi \int_{-\infty}^{\infty} |F_a(r)|^2 r dr \int_{-\infty}^{\infty} |F_b(r)|^2 r dr} \tag{16a}$$

$$f_{abcd} = \frac{\int_{-\infty}^{\infty} F_a^*(r) F_b^*(r) F_c(r) F_d(r) r dr}{2\pi \sqrt{\int_{-\infty}^{\infty} |F_a(r)|^2 r dr \int_{-\infty}^{\infty} |F_b(r)|^2 r dr \int_{-\infty}^{\infty} |F_c(r)|^2 r dr \int_{-\infty}^{\infty} |F_d(r)|^2 r dr}} \tag{16b}$$

Here, A_i ($i = p$ for pump, i for idler and s for signal field) are the amplitudes of waves, α_{1i} and α_{2i} ($i = p$ for pump, i for idler and s for signal field) denote the linear and nonlinear absorption coefficients at the corresponding wavelengths (88). The parameters, f_{ab} (a, b for the pump, idler and signal fields) symbolize the effective overlap between any two modes, the a and b modes: f_{aa} describe the self-phase modulations and f_{ab} are for the cross-phase modulations as defined in Eq. 16(a) (Agarawal, 2008). The next parameters, f_{abcd} (a, b are for the pump, idler, and the signal fields) symbolize overlap among the four modes (Agarawal, 2008). The function $F_a(r)$ represents the modal field profile as a function of the radial coordinate for the ath wave.

A pump power of 1 kW and an idler seed power of 20 W are injected into the fiber, and their variation with the fiber length L are calculated. Due to the coherence length, $L_c \sim 1/\Delta k$, the power of each of the three fields varies periodically as shown in Figure 17(b). First, both idler and signal power increase at the cost of the pump power and reach the maximum at the length equal to the coherence length in the fiber. The corresponding length of the fiber at which the idler and signal power reach the maximum is also called the optimized length (L_0). With the increasing input pump power, the optimized length of the fiber decreases. An optimized length of 6.5 m has been found at 9kW of pump power which produces a maximum signal at 9.33 THz, while optimum signal output at 9.14 THz is obtained at the optimized length of 37.8 m using 1 kW pump power. The variation of the optimized length, the power amplification factor (PAF) and the THz generation efficiency (η) have been studied with respect to the increasing pump power by keeping the idler seed power constant at 20 W, and the increasing idler seed power by keeping the pump power constant at 1kW. The corresponding results are presented in Figures 18(a) and 18(b), respectively. Here, PAF is the ratio of the THz signal power to the fixed input idler seed power of 20 W and η is the ratio of the THz signal power to the fixed input pump power of 1kW. For these calculations, three sets of absorption parameters have been used as indicated in Figure 18(a). Clearly, at constant idler power of 20 W, the L_0 decreases, PAF increases, and the efficiency attains a saturation value with the increasing pump power as shown in Figure 18(a). On the other hand, at constant pump power of 1kW, the L_0 decreases rather slowly than the above case, PAF decreases, and the efficiency attains a saturation value with the increasing idler seed power as shown in Figure 18(b). It can also be noted from these results that higher the nonlinear absorption in the fiber is, higher is the optimized fiber length and lower are the PAF and η.

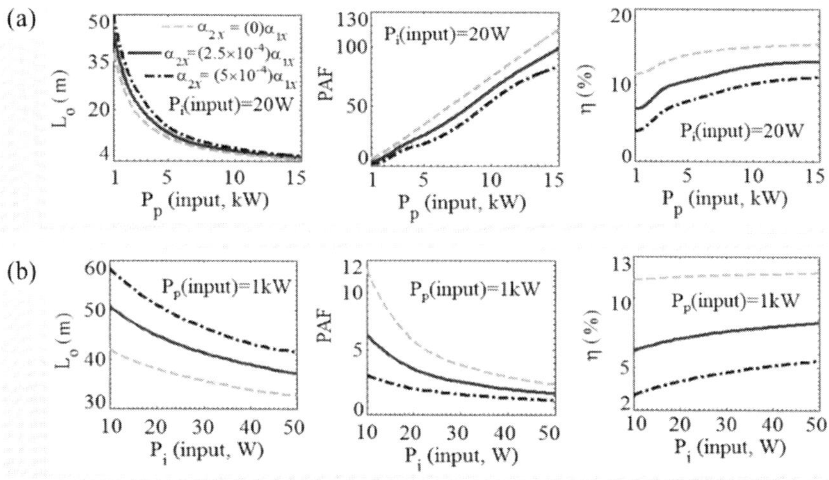

Figure 18. Nonlinear absorption (α_{2x}) dependent performance for THz signal generation and guidance in microstructured fiber schematic as shown in Figure 15(a). (a) Input pump power-dependent results for the optimized length (L_o), power amplification factor (PAF), and efficiency (η) at fixed input idler power of 20 W. (b) Input idler power-dependent results for the same parameters as above and at a fixed input pump power of 1 kW.

The effect on the fiber transmission characteristics due to bending have also been determined. The bending loss in dB/km is an important characteristic of any fiber. It has been calculated for the PCF design of Figure 15(a) using the vector mode power overlap integral method, where the integration of the transmitted power is performed by taking the transverse effective refractive index distribution in the bent fiber (Neumann, 1998; Baggett, 2003). Because of finite bending with the bending radius R_b, the effective refractive index profile from its original form, $n_0(x,y)$ for the straight fiber gets shifted to a new form, $n(x,y)$. Figure 19(a) presents the results for the shift in the modal profile with respect to two values of the bending radius relative to the case of straight fiber ($R_b = \infty$). Clearly, up to a bending radius of 1 cm, there is a minimal shift in the modal profile, however, for lower values of bending radii, there is a large shift obtained for the modal profile. The bending loss in the fiber at a frequency of = 9.14 THz as a function of the bending radius has been determined and is presented in Figure 19(b). The loss is negligible for bending radius larger than 2 cm. Therefore, a mandrel with radius of 3 cm can be used for wrapping the PCF around it while its being used for the generation of the THz radiation by FWM.

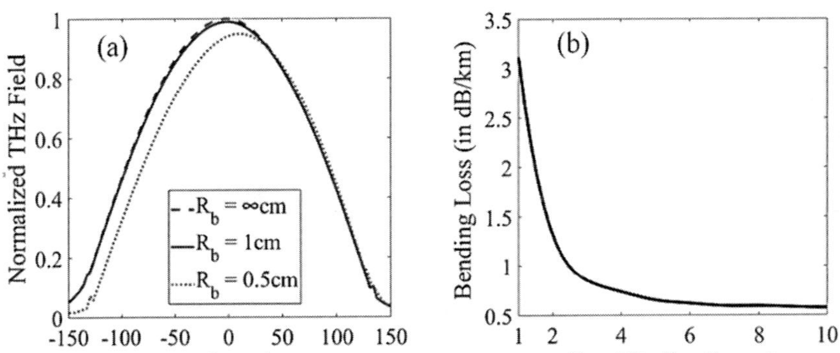

Figure 19. (a) THz signal modal field profiles at 9.14 THz for different bend radii of the PCF schematic shown in Figure 15(a). (b) Variation of the bend loss coefficient (α_b) with bend radius for the generated THz signal as calculated by using the overlap vector integral method.

Broad-Band Nonlinear PCF THz Source

In the previous section, we have seen a monoatomic topology-based PCF design for efficient THz generation in a narrow frequency range (Barh, 2015; Brothers, 1994; Demas, 2017; Jauregui, 2012; Kumar, Optics communication, 2020). Here, we describe the design and properties of a diatomic topology-based PCF design. Two designs of the PCF configurations in the Teflon-1 host material have been considered, namely, conventional solid-core PCF and diatomic hexagonal ring-based solid core PCF (Kumar, *Conference on Lasers and Electro-Optics Europe & European Quantum Electronics Conference, 2019*). Transverse cross-sections corresponding to these solid-core PCFs are shown schematically in Figures 20(a) and 20(b), respectively. The clad consists of a triangular lattice of air columns in the first type of the conventional solid-core PCF design, while alternate hexagonal rings of Teflon-2 and air columns have been considered for the second type of the solid core PCF design. In both the schematics (Figure 20), the number of the rings in the cross-section has been considered to be the same. The diatomic lattice-based PCF design helps to achieve the generation of THz supercontinuum by excitation with an ultrashort THz pulse centered at the wavelength of 225μm (frequency of 1.33THz) with a pulse duration of 5ps and its peak power of 200W. The electric field of THz pulse can be expressed as $E(t) = Ae^{-t^2/\tau_0^2}$, where A is the electric field amplitude and τ_0 is the pulse duration. As shown in Figures 20(a) and 20(b), the pitch, Λ, is 300μm. The diameter (d) of inclusions in the cladding is varied from 0.5 Λ to 0.9Λ to tailor the spectral dispersion profiles of the PCF as shown in Figures 21(a) and 21(b).

The modal characteristics for the two solid-core PCF designs have been calculated using full vector beam propagation method. A comparison between the THz dispersion profiles of the two PCF designs in Figures 21(a) and 21(b) suggests that the diatomic PCF dispersion is not only lesser, but it is also anomalous in a large frequency range, as compared to that of the monoatomic PCF dispersion. Therefore, the prior is suitable for the generation of THz supercontinuum and accordingly, optimization in the structure has been done for efficient generation of broadband THz radiation from it.

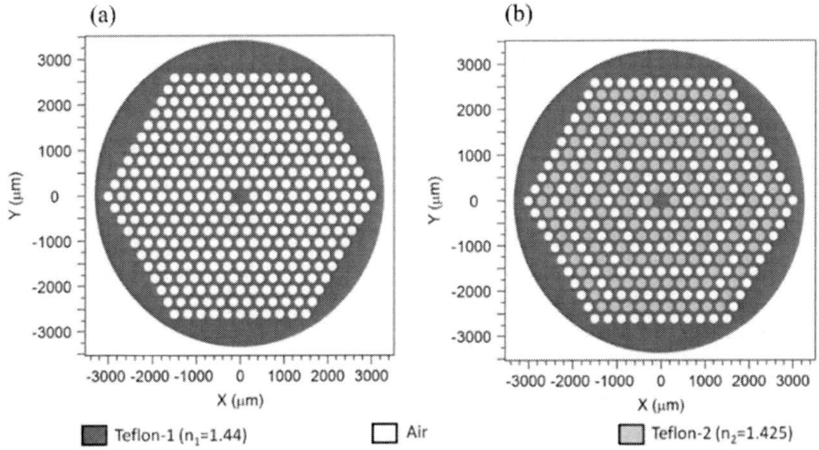

Figure 20. Cross-sectional views of a conventional solid-core PCF and diatomic hexagonal rings-based solid-core PCF. The host material is considered to be Teflon-1. (a) The clad consists of a triangular lattice of air columns in conventional solid-core PCF and (b) the clad consists of alternate hexagonal rings of Teflon-2 and air columns. The number of rings has been considered same in the both the schematics.

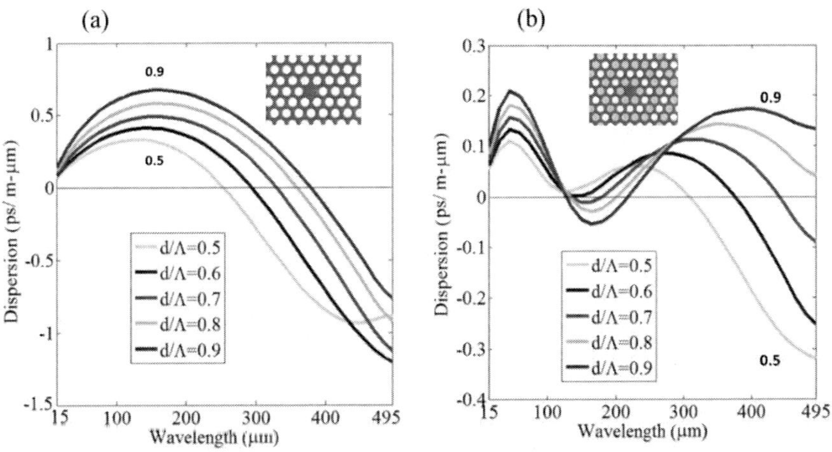

Figure 21. Spectral dispersion characteristics of (a) conventional solid-core PCF and (b) diatomic hexagonal rings based solid core PCF as shown in Figures 20(a) and 20(b), respectively, evaluated using full vector beam propagation method for different diameter to pitch ratios (*d*/*Λ*) as indicated.

Like in any other nonlinear application, the modal effective area of the PCF is expressed as the following,

$$A_{eff} = \frac{\left(\int |E|^2 dS\right)^2}{\int |E|^4 dS} = \frac{\left(\int I dS\right)^2}{\int I^2 dS} \tag{17}$$

where, E is the associated electric field, and dS is the differential surface area. The integration is done over the entire cross-section. The spectral variation of the modal area for different d/Λ ratios in the PCF are shown in Figure 22(a). After calculating the modal effective area, the nonlinear coefficient (γ) at a frequency ω has been calculated using the formula,

$$\gamma = \frac{n_2 \omega}{c A_{eff}} \tag{18}$$

Here, the other symbols have their usual meaning, i.e., n_2 is the nonlinear index of refraction and c is the speed of light in vacuum. The spectral variation of the nonlinear coefficient and its angular frequency derivative for a few d/Λ values are presented in Figure 22(b) and 22(c), respectively. These results are then used in the usual manner, in a nonlinear Schrodinger's equation for determining the THz generation from the PCF.

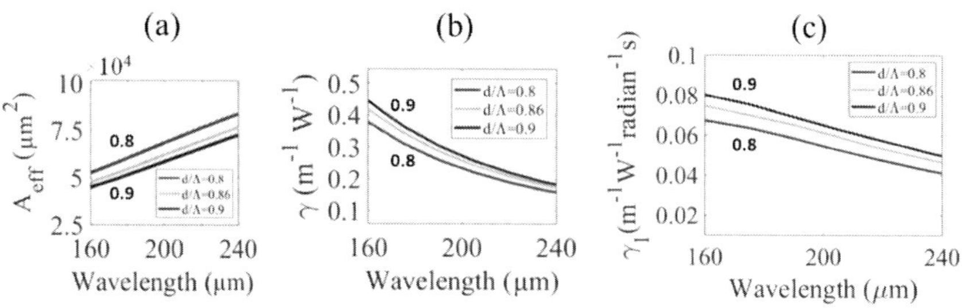

Figure 22. Spectral variation of (a) effective mode area, (b) nonlinear coefficient, and (c) angular frequency derivative of the nonlinear coefficient for the diatomic hexagonal rings-based solid core PCF depicted in Figure 20(b).

The generalized nonlinear Schrödinger equation (GNLSE) is solved by using the split-step method (Agarawal, 2007), which numerically calculates the THz pulse evolution in the diatomic PCF.

$$\frac{\partial A}{\partial Z}+\frac{1}{2}(\alpha(\omega_0)+i\alpha_1(\omega_0)\frac{\partial}{\partial t})A+\beta_1\frac{\partial A}{\partial t}+\frac{i\beta_2}{2}\frac{\partial^2 A}{\partial t^2}-\frac{\beta_3}{6}\frac{\partial^3 A}{\partial t^3}$$
$$=i(\gamma(\omega_0)+i\gamma_1(\omega_0)\frac{\partial}{\partial t})(A(z,t)\int_0^\infty R(t')\left|A(z,t-t')\right|^2 dt') \tag{19}$$

In the equation above, $\alpha(\omega_0)$ is the absorption coefficient at the center frequency of ω_0, $\alpha_1(\omega_0)$ is the first derivative of the absorption coefficient with respect to frequency, A is the THz amplitude, and β_i (i = 1, 2, etc.) denote the various dispersion coefficients of the PCF. The factor, $R(t')$ is the combination of the instantaneous electronic effect due to self-phase modulation, its higher-order effects, and the Raman factor (f_r), and is given by the expression below.

$$R(t')=(1-f_r)\delta(t'-t_e)+f_r h_r(t') \tag{20}$$

Here, $\delta(t'-t_e)$ represents the instantaneous electronic effect due to self-phase modulation and its higher-order effects and $h_r(t')$ represents a unit step-function to account for the Raman. The latter is given by the following expression,

$$h_r(t')=\frac{\tau_1^2+\tau_2^2}{\tau_1\tau_2^2}\exp(-t'/\tau_1)\sin(t'/\tau_2) \tag{21}$$

where, τ_1 and τ_2 are two characteristic response time-constants for Teflon material arising from damping and resonant scattering in the Raman susceptibility.

Table 2. Fiber properties and input pulse characteristics suitable for supercontinuum generation from a PCF with d/Λ = 0.9

Dispersion coefficients β_2 and β_3	-11.83×10^{-5} ps^2/m, 8.103×10^{-7} ps^4/m
Fiber Length	L = 1.5m
Nonlinear Coefficient (Υ) (Agarawal, 2007)	$n_2 = 3.19\times10^{-14}$ m^2/W, $\Upsilon = 0.21$ m^{-1}W^{-1}
Peak Power	200 W
Pulse Duration	τ_0 = 5 ps
Raman factor	f_R = 0.45
Time Scales (Kumar, 2020)	τ_1 = 100ps, τ_2 = 330ps
Loss (R J Chase company, PTFE)	0.017cm^{-1}

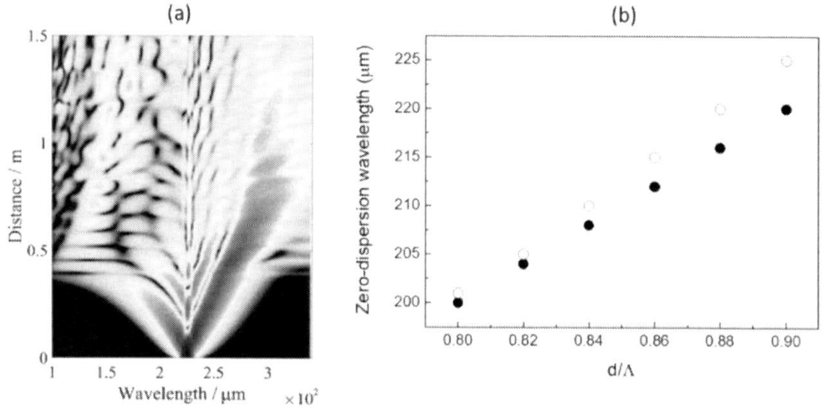

Figure 23. (a) THz supercontinuum from the solid core PCF as shown in Figure 20(b) for a pump wavelength of 225 µm, fiber geometry of d/Λ = 0.9 and other parameters given in Table 2. (b) Zero-dispersion wavelength (closed symbols) for different geometries (*d*/*Λ*) of the solid core PCF for adequate supercontinuum generation at THz frequencies. The open symbols represent the excitation pump-wavelength within the acceptance range dictated by the dispersion behaviour as shown in Figure 21(b).

The PCF characteristics and those of the excitation pulse used in this study are presented in Table 2. The supercontinuum evolution in the proposed structure with *d*/*Λ* = 0.9 is shown in Figure 23(a), where spectral evolutions of the input pump pulse are indicated. The spectral broadening at the small length of the PCF occurs due to self-phase modulation. Beyond a certain length, the Raman effect causes broadening of the spectrum. The dispersion effects are more severe at the low wavelengths. The present results suggest that the efficiency of THz supercontinuum generation is better from a diatomic lattice-based PCF as compared to that from a conventional PCIGF (Kumar, *Conference on Lasers and Electro-Optics Europe & European Quantum Electronics Conference, 2019*). By adjusting the *d*/*Λ* ratio, it is possible to optimize the output at different excitation wavelengths according to the zero-dispersion point in the spectral dispersion profile of the PCF. The excitation wavelength should be in the anomalous dispersion region (longer than the zero-dispersion wavelength) and close to the zero-dispersion point such that the low dispersion part contributes better to the spectral broadening by self-phase modulation, soliton generation and soliton fission, and the Raman shifting of the solitons to longer wavelengths. From the calculated dispersion profiles of the PCF, the zero-dispersion wavelengths and corresponding excitation wavelengths have been shown in Figure 23(b) for a quick reference.

Conclusion

Different designs and other properties of microstructured fibers have been discussed in this chapter which, make PCFs suitable as sources and other applications in the THz frequency range. Optimization of the designs in either the core or the clad allows tuning of the THz properties as per the needs for specific applications. Particularly, we have presented those to be used as THz sources of narrow and broadband radiation, and polarization devices based on various popular polymeric materials as the host as well as the micro-inclusions in the core or the clad configurations.

Acknowledgements

The authors thank IIT Delhi for support in carrying out the research work. SK acknowledges funding support from Joint Advanced Technology Center for support through the research project, EMDTERA#5. VK thanks University Grants Commission for Senior Research Fellowship.

References

Angelo, F. D., Mics, Z., Bonn, M. and Turchinovich, D. (2014). Ultra-broadband terahertz time-domain spectroscopy of common polymers using THz air photonics. *Optics express*, 12475.

Abina, A., Puc, U., Jeglic, A. and Zidansek, A. (2016). Structural characterization of thermal building insulation materials using terahertz spectroscopy and terahertz pulse imaging. *NDT&E international*, 11-18.

Agarawal, A., Kejalakshmy, N., Uthman, M., Rahman, B. M. A., Kumar, A. and Grattan, K. T. V. (2012). Ultra low bending loss equiangular spiral photonic crystal fibers in the terahertz regime. *Advancement in physics advances*, 022140.

Arora, V., Varshney, R. K. and Kumar, V. 2018 "Enhancement of Photonic Band-gap for Terahertz," *Optics and Photonics: Theory and Computational Techniques workshop at I.I.T. Roorkee, India* during March 17-18, 2018.

Arora, V., Kumar, V. and Varshney, R. K. 2018. "Deformed core high birefringent photonic crystal fiber for THz guidance," *International Conference of Fiber Optics and Photonics (Photonics-2018 conference)* Paper No. SP153, ISBN 978-93-88653-41-1.

Arora, V., Kumar, V. and Varshney R. K. 2018. "High Birefringent Photonic Crystal Fibers for THz Guidance," *XLII Optics Society of India, International Symposium of Optics (OSI-ISO symposium) held at I.I.T. Kanpur, India during September 19-22.*

Agarawal, G. P. 2007. *Nonlinear Fiber optics*, Elsevier, Academic Press.

Baggett, J. C., Monro, T. M., Furusawa, K., Finnazi, V. and Richardson, D. J., (2003). "Understanding bending losses in holey optical fibers. *Optics communication*, 317-335.

Bogue, R. (2018). Sensing of terahertz radiation: a review of recent progress," *Sensor Review*, 216-222.

Barh, A., Varshney, R. K., Agrawal, G. P., Rahman, B. M. A. and Pal, B. P. (2015). Plastic fiber design for THz generation through wavelength translation," *Optics letters*, 2107-2110.

Barh, A., Pal, B. P., Agrawal, G. P., Varshney, R. K. and Rahman, B. M. A. (2016). Specialty Fibers for Terahertz Generation and Transmission: A Review. *IEEE journal of selected topics in quantum electronics*, 365-379.

Blake, G. A. 2008. "Terahertz Spectroscopy in the Lab and at Telescopes" *33rd International Conference on Infrared, Millimeter and Terahertz Waves,* DOI-10.1109/ICIMW.2008.4665863

Bouwmans, G., Luan, F., Knight, J. C., Russell, P. St. J., Farr, L., Mangan, B. J. and Sabert, H. (2003). Properties of a hollow-core photonic bandgap fiber at 850 nm wavelength. *Optics express*, 1613-1620.

Burrow, J. A., Yahiaoui, R., Sarangan, A., Agha, I., Mathews, J. and Searles, T. A. (2017). Polarization-dependent electromagnetic responses of ultrathin and highly flexible asymmetric terahertz metasurfaces. *Optics express*, 32540-32549.

Boyd, R. W. 2008. *Nonlinear Optics*, Elsevier, Academic Press.

Brothers, L. R., Lee, D. and Wong, N. C. (1994). Terahertz optical frequency comb generation and phase locking of an optical parametric oscillator at 665 GHz. *Optics letters*, 245-247.

Chen, J., Chen, Y., Zhao, H., Baastians, G. J. and Zhang, X. C. (2007). Absorption coefficients of selected explosives and related compounds in the range of 0.1-2.8 THz. *Optics express*, 12060.

Chen, H., Yan, G., Forsberg, E. and He, S., (2016). Terahertz polarization splitter based on a dual-elliptical-core polymer fiber. *Applied optics*, 6236-6242.

Cosentino, A. (2016). Terahertz and Cultural Heritage Science: Examination of Art and Archaeology," *Technologies*, 6.

Dragoman, D. and Dragoman, M. (2004). Terahertz fields and applications. *Progress of quantum optics,* 1-66.

Demas, J., Prabhakar, G., He, T. and Ramachandran, S. (2017). Wavelength-agile high-power sources via four-wave mixing in higher-order fiber modes. *Optics express*, 7455-7464.

Exter, M. V., Fattinger, C. and Grischkowsky, D. (1989). Terahertz time-domain spectroscopy of water vapor," *Optics letters*, 14, 1128-1130.

Gallot, G., Jamison, S. P., McGowan, R. W. and Grischkowsky, D. (2000). Terahertz waveguides. *Journal of optical society of America B*, 851-863.

Guo, F., Pandey, C., Wang, C., Nie, T., Wen, L., Zhao, W., Miao, J., Wang, L. and Wu, X. (2020). Generation of highly efficient terahertz radiation in ferromagnetic heterostructures and its application in spintronic terahertz emission microscopy (STEM). *O.S.A. continuum*, 893-902.

Hangyo, M. (2015). Development and future aspects of terahertz technology. *Japanese Journal of applied physics*, 120101.

Hafez, H. A., Chai, X., Ibrahim, A., Mondal, S., Ferachou, D., Robagnol, X. and Ozagi, T. (2016). Intense terahertz radiation and their applications. *Journal of optics*, 093004.

Han, H., Park, H., Cho, M. and Kim, J. (2002). Terahertz pulse propagation in a plastic photonic crystal fiber *Applied physics letters*, 2634.

Healy, N., Sparks, J. R., He, R. R., Sazio, P. J. A., Badding, J. V. and Peacock, A. C. (2011). High index contrast semiconductor ARROW and hybrid ARROW fibers. *Optics express,* 10979-10985.

Hassani, A., Dupuis, A., and Skorobogatiy, M. (2008). Surface-plasmon-resonance-like fiber-based sensor at terahertz frequencies. *Journal of optical society of America B*, 1771-1775.

Hazhorova, A., Markov, A., Ng, A., Chinnappan, R., Zourob, M., and Skorobogatiy, M. (2012). Label-free bacteria detection using evanescent mode of a suspended form terahertz fiber. *Optics express*, 5344-5355.

Hidayat, A., Koch, B., Zhang, H., Mirvoda, V., Lichtinger, M., Sandel, D. and Noé, R. (2008). High-speed endless optical polarization stabilization using calibrated waveplates and field-programmable gate array-based digital controller. *Optics express*, 18984-18991.

Hibbard, M. T., Lake, D. S., Johasson, N. A. B., Thomson, T., Jamison, S. P. and Graham D. M. (2019). Magnetic field tailoring of the terahertz polarization from a spintronic source. *Applied physics letters*, 031101.

Islam, M. S., Sultana, J., Faisal, M., Islam, M. R., Dinovitser, A., Ng, B. W. H. and Abbott, D. (2018). A modified hexagonal photonic crystal fiber for terahertz applications," *Optical materials*, 336-339.

Jauregui, C., Steinmetz, A., Limpert, J. and Tünnermann A. (2012). High-power efficient generation of visible and mid-infrared radiation exploiting four-wave-mixing in optical fibers. *Optics express*, 24957-24965.

Jazbinsek, M., Puc, U., Abina, A. and Zidansek, A. (2019). Organic crystals for THz photonics. *Applied science*, 882.

Jena, J., Wassin, S., Bezuidenhout, L., Doucouré, M. and Gibbon, T. (2020). Polarization-based optical fiber acoustic sensor for geological applications. *Journal of optical society of America B*, A147-A153.

Khaleque, A. and Hattori, H. T. (2015). Ultra-broadband and compact polarization splitter based on gold filled dual core photonic crystal fiber. *Journal of applied physics*, 143101.

Kozlov, S. A., Drozlov, A. A., Dolgleva, K., and Boyd, R. W. 2015. "Nonlinear refractive index for crystals at terahertz frequencies" 40th *International conference of infrared, millimeter and terahertz waves* (IRMMW-THz).

Kumar, V., Varshney, R. K. and Kumar, S. (2020). Two interpenetrating hexagonal lattice in clad photonic crystal fiber for THz sensing," *Frontiers in Optics/Laser Science,* Paper No. FM7B.5. DOI: 10.1364/FIO.2020.FM7B.5

Kao, T. Y., Reno, J. L. and Hu, Q. (2017). Amplifiers of Free-Space Terahertz Radiation. *Optica*, 713-716.

Kim, S., Kee, C.K. and Lee J. M. (2007). Single-mode Condition and Dispersion of Terahertz Photonic Crystal Fiber. *Journal of optical society of Korea*, 97-100.

Kumar, V., Varshney, R. K. and Kumar, S. 2018. "Polyethylene based Photonic Band Gap Fiber for Terahertz Applications" *Optics and Photonics: Theory and Computational Techniques workshop at I.I.T. Roorkee, India* during March 17-18.

Kumar, V., Varshney, R. K. and Kumar, S. 2018. "Photonic Band Gap Fibers for Terahertz Applications," *Recent Trends in Advances in Physics, an annual symposium of Department of Physics, I.I.T. Delhi, India*, held during March 20-21.

Kumar, V., Varshney, R. K. and Kumar, S. 2019 "Broadband and flat THz dispersion in biatomic photonic band gap fibers" *Workshop on Recent Advances in Photonics, Poster session No. pp. 1-3, (WRAP 2019).* DOI: 10.1109/WRAP47485.2019.9013698.

Kumar, V., Varshney, R. K. and Kumar S. 2018. "Near zero-dispersion broadband THz guidance in microstructured fibers," *Photonics 2018 conference, held during December 12 -15, 2018 at I.I.T. Delhi, India,* Paper No. SD1-C3, ISBN 978-93-88653-41-1 (2018).

Kumar, V., Varshney, R. K. and Kumar S. 2021. "Broadband low dispersion THz photonic crystal fiber" in *Optics and Photonics: Theory and Computational Techniques workshop held online and conducted by I.I.T. Delhi.*

Kumar, V., Varshney, R. K. and Kumar S. 2019. "Near Zero Flat Dispersion by Hexagonal Patterned Unit Cell in THz Photonic Crystal Fiber," *Springer Proceedings in Physics,* vol. 258 (ICOL 2019). DOI: 10.1007/978-981-15-9259-1_203

Koch, B., Noé, R., Sandel, D. and Mirvoda, V. (2014). Versatile endless optical polarization controller/tracker/demultiplexer. *Optics express*, 8259-8276.

Kumar, V., Varshney, R. K. and Kumar, S. (2020). Design of a compact and broadband terahertz polarization splitter based on gradient dual-core photonic crystal fiber. *Applied optics*, 1974-1979.

Kumar, V., Varshney, R. K. and Kumar, S. (2021). Compact and efficient polarization splitter based on dual core microstructured fiber for THz applications. *Results in optics*, 100094.

Kumar, V., Varshney, R. K. and Kumar, S. (2020). Terahertz generation by four-wave mixing and guidance in diatomic Teflon photonic crystal fibers. *Optics communication*, 124460.

Kumar, V., Varshney, R. K. and Kumar S. (2019) "Terahertz Supercontinuum in the Biatomic Photonic Crystal Fiber," *Conference on Lasers and Electro-Optics Europe & European Quantum Electronics Conference (CLEO/Europe-EQEC), Munich,* Germany, pp. 1-1, DOI: 10.1109/CLEOE-EQEC.2019.8872721.

Li, S., Zhang, H., Hou, Y., Bai, J., Liu W. and Chang, S. (2017). Terahertz polarization splitter based on orthogonal microstructure dual-core photonic crystal fiber. *Applied optics*, 3305-3310.

Li, J. S., Liu, H. and Zhang, L. (2014). Terahertz wave polarization beam splitter using a cascaded multimode interference structure. *Applied optics*, 5024-5028.

Liu, C., Wang, L., Wang, F., Xu, C., Liu, Q., Liu, W., Yang, L., Li, X., Sun, T. and Chu, P. K. (2019). Tunable single-polarization bimetal-coated and liquid-filled photonic crystal fiber filter based on surface plasmon resonance. *Applied optics*, 6308-6314.

Menyuk, C. R. and Marks, B. S. (2006). Interaction of Polarization Mode Dispersion and Nonlinearity in Optical Fiber Transmission Systems. *Journal of lightwave technology*, 2806.

Morris, C. M., Aguilar, R. V., Stier, A. V. and Armitage, N. P. 2012. "Polarization modulation time-domain terahertz polarimetry," in *Imaging and Applied Optics Technical Papers*, O.S.A. Technical Digest (online) (Optical Society of America), paper SW4C.2.

Mittleman D. L., Gupta M., Neelamani R., Baraniuk R. G., Rudd J. V., Koch M. (1999). Recent advances in terahertz imaging," *Applied physics B*, 1085-1094.

Markov, A. and Skorobogatiy, M. (2013). Two-wire terahertz fibers with porous dielectric support. *Optics express*, 12728-12743.

Mazhorova A., Zourob M., and Skorobogatiy M. (2011). Suspended core polyethylene fiber for bio-sensing applications in the terahertz region," in *Advanced Photonics*, O.S.A. Technical Digest (CD) (Optical Society of America, 2011), paper STuC3.

Mihály, J., Sterkel, S., Ortner, H. M., Kocsis, L., Hajba, L., Furdyga, É. and Mink, J. (2006). FTIR and FT-Raman Spectroscopic Study on Polymer Based High Pressure Digestion Vessels. *Original scientific paper*, 497-501.

Neumann, E. G. 1998. *Single-Mode Fibers Fundamentals*, Springer-Verlag.

Pakarzadeh, H., Rezaei, S. M. and Naamroody, L. (2019). Hollow-core photonic crystal fibers for efficient terahertz transmission. *Optics communications*, 81-88.

Pawar, A. Y., Sonawane, D. D., Erande, K. B., and Derle, D. V. (2013). Terahertz technology and its applications. *Drug invention today*, 157-163.

Pedrola, L. 2016. *Beam propagation methods for the design of optical waveguide devices*, Wiley.

Pysz, D., Kujawa, I., Stępień, R., Klimczak, M., Filipkowski, A., Franczyk, M., Kociszewski, L., Buźniak, J., Haraśny, K. and Buczyński, R. (2016). Stack and draw fabrication of soft glass microstructured fiber optics. *Boulletin of the Polish academy of science*, 667-682.

Park, J. M., Lee, S. G., Park, H. R. and Lee, M. H. (2010). High-efficiency polarization beam splitter based on a self-collimating photonic crystal. *Journal of optical society of America B*, 2247-2254.

Qasymeh, M. (2012). Terahertz Generation in an Electrically Biased Optical Fiber: A Theoretical Investigation," *International journal of optics*, 1-6.

Rozé, M., Ung, B., Mazhorova, A., Walther, M. and Skorobogatiy, M. (2011). Suspended core subwavelength fibers: towards practical designs for low-loss terahertz guidance. *Optics express*, 9127-9138.

Russell, P. St. J. (2006). Photonic-Crystal Fibers. *Journal of lightwave technology*, 4729-4749.

R J Chase company, Inc. 4000 Tara Court Union City, CA, 94587, Whitford, SPI member (http://www.rjchase.com/ptfe_handbook.pdf)

Rahman B. M. A. and Agrawal, A. 2013. *Finite Element Modeling Methods for Photonics*, Artech House.

Sesay, M., Jin, X. and Ouyang, Z. (2013). Design of polarization beam splitter based on coupled rods in a square-lattice photonic crystal. *Journal of optical society of America B*, 2043-2047.

Sharma, M., Borogohain, N. and Konar, S. (2013). Index Guiding Photonic Crystal Fibers With Large Birefringence and Walk-Off. *Journal of lightwave technology*, 3339-3344.

Song, H. J. and Nagatsuma, T. (2011). Present and future of terahertz communications. *IEEE Transactions on Terahertz Science and Technology*, 256-263.

Sørgård, T., Song, S., Vullum, P. E., Kores, C., Mølster, K. M., Laurell, F., Hawkins, T., Ballato, J., Österberg, U. L. and Gibson, U. J. (2020). Broadband infrared and THz transmitting silicon core optical fiber. *Optics material express*, 2491-2499.

Suizu, K. and Kawase, K. (2007). Terahertz-wave generation in a conventional optical fiber," *Optics letters*, 2990-2992.

Sukhoivanov, I. A. and Guryev, I. V. 2009. *Photonic Crystals- Physics and Practical Modeling*, Springer-Berlin Heidelberg.

Tang, P. R., Li, J., Du, L. H., Liu, Q., Peng, Q. X., Zhao, J. H., Zhu, B., Li, Z. R., and Zhu, L. G. (2018). Ultrasensitive specific terahertz sensor based on tunable plasmon induced transparency of a graphene micro-ribbon array structure. *Optics express*, 30655-30666.

Wang, J., Pei, L., Weng, S., Wu, L., Li, J. and Ning, T. (2018). Ultrashort polarization beam splitter based on liquid filled dual-core photonic crystal fiber *Applied optics*, 3847-3852.

Wu, Z., Shi, Z., Xia, H., Zhou, X., Deng, Q., Huang, J., Jiang, X. and Wu W. (2018). Design of Highly Birefringent and Low-Loss Oligoporous-Core THz Photonic Crystal Fiber With Single Circular Air-Hole Unit," *Optics express*, 6.

Xie, L., Yao, Y. and Ying, Y. (2014). The applications of terahertz spectroscopy: to protein detection: A review, *Applied spectroscopy reviews*, 448-461.

Xu, K., Bayati, E., Oguchi, K., Watanabe, S., Winebrenner, D. P and Arbab, M. H. (2020). Terahertz time-domain polarimetry (THz-TDP) based on the spinning E-O sampling technique: determination of precision and calibration. *Optics express*, 13482-13496.

Yan, F., Li, L., Wang, R., Tian, H., Liu, J., Tian, F., and Zhang, J. (2019). Ultrasensitive Tunable Terahertz Sensor with Graphene Plasmonic Grating. *Journal of lightwave technology*, 1103-1112.

Yang, X., Zhao, X., Yang, K., Liu, Y., Liu, Y., Fu, W. and Luo, Y. (2016). Biomedical applications of terahertz spectroscopy and imaging," *Trends of biotechnology*, 810.

Yang, T., Ding, C., Ziolkowski, R. W. and Guo, Y. J. (2018). Circular hole E.N.Z. photonic crystal fibers exhibit high birefringence, *Optics express*, 17264-17278.

Zhong, S. and Sheng, L. (2015). Terahertz wave polarization splitter using full bandgap photonic crystal. *Journal of infrared, millimeter and terahertz waves*, 255-261.

Zhou, P. and Fan, D. (2011). Terahertz-wave generation by surface-emitted four-wave mixing in optical fiber," *Chinese optics letters*, 051902-1-4.

Chapter 7

The Fiber Fuse Effect in Hollow-Core Photonic Crystal Fibers

Y. Shuto*, PhD

Ofra Project, Iruma, Japan

Abstract

Recently great interest has been aroused by the development of photonic crystal fibers (PCFs). Hollow-core photonic crystal fibers (HC-PCFs) can guide light inside the hollow core by photonic bandgap or inhibited-coupling mechanisms. HC-PCFs exhibit low propagation losses at 1.064 μm and the 1.51-1.60 μm range. The HC-PCFs are now implemented as an important tool of laser beam delivery in place of traditional single mode optical fibers. One of the problems arising from high-power injection in the optical fiber is the probability of triggering the fiber fuse effect. This effect is related to a heat conduction process in optical fibers. The unsteady-state thermal conduction process in several HC-PCFs was studied theoretically by the explicit finite-difference method using the thermochemical SiO_x production model. For heat conduction analysis, the complicated inner structure of HC-PCFs was simplified using the model composed of silica-ring and air-hole layers. In a kagome-lattice HC-PCF, the inclusion of metallic impurities in the silica struts was responsible for fiber fuse initiation. The calculated velocities of fiber fuse propagation in the kagome-kattice PCF were in fair agreement with the experimental values. Furthermore, the calculated velocities of fiber fuse propagation in two types of polymer-coated revolver fibers were in fair

* Corresponding Author's Email: ofra@tuba.ocn.ne.jp.

In: Photonic Crystal Fibers (PCFs)
Editor: Mário F. S. Ferreira
ISBN: 979-8-88697-618-2

agreement with the experimental values. If the polymer coating of the revolver fiber is removed, the Fresnel reflection at the outer surface of the support tube occurs and the back-reflected light wave is incident upon the silica capillaries and the hollow core. To clarify the in-phase condition, we estimated phase changes of optical routes in the uncoated revolver fiber. It was found that the reflected waves from the outer surfaces of the silica capillary and support tube are in phase at the core-capillary boundary, and they are mutually enhanced as a consequence of the constructive interference. As a result, the power in the hollow core and the silica capillary was improved for the uncoated revolver fiber compared with the polymer-coated one. The calculated velocity of fiber fuse propagation in the uncoated revolver fiber was in fair agreement with the experimental value.

Keywords: photonic crystal fiber, fiber fuse phenomenon, finite-difference technique

Introduction

Recently great interest has been aroused by the development of photonic crystal fibers (PCFs) [1-98]. In PCFs, a microstructured or photonic crystal material forms the fiber cladding. The core in which the light propagates is formed by omitting one or more periods of the photonic crystal (leaving an air core) or by filling in one or more air holes (forming a solid core). The former is called a hollow-core PCF (HC-PCF) and the latter is called a holey fiber (HF) or a microstructured optical fiber (MOF). These PCFs have been under intensive study because they offer design flexibility in the control of modal properties. The PCFs have some extraordinary properties, such as a wide single-mode wavelength range [2, 3, 5] and large group velocity dispersion (GVD) and/or flattened GVD [4, 12, 13, 24, 31, 32]. Furthermore, the PCFs show some superior characteristics, such as an ultralow transmission loss [33, 37, 40, 42, 88, 90, 95, 96], significantly enhanced effective nonlinearity in holey fibers [11], and greatly reduced effective nonlinearity in hollow-core fibers [29, 54].

Hollow-core photonic crystal fibers (HC-PCFs) can guide light inside the hollow core by photonic bandgap (PBG) [27] or inhibited-coupling (IC) mechanisms [53]. The confinement loss in IC-guiding HC-PCFs, represented by kagome-lattice HC-PCFs [21], has been reduced from the 1 dB/m level to

8.5-70 dB/km with the introduction of a fiber core with a hypocycloid-like shape [66, 67, 84, 87]. The role of the hypocycloid shape in decreasing the confinement loss was explained by an enhancement of IC via a dramatic decrease in the spatial overlap between the slow-varying transverse-phase core modes and the fast-varying transverse cladding modes [62, 99].

Furthermore, the relationship between the negative curvature and the confinement loss in the HC-PCF family has been experimentally and theoretically investigated by several research institutes [63, 64, 69, 70, 72-81, 85, 88-91, 95, 96, 99, 100]. Pryamikov et al. fabricated a novel hollow-core PCF with a negative curvature of the core boundary and a cladding consisting of one row of capillaries [63]. This fiber is often called an antiresonant fiber [72-77, 82, 88], a negative curvature fiber [64, 69, 70, 81, 83, 100], or a revolver fiber [80, 89].

Recently, Sakr et al. reported the fabrication of the nested antiresonant nodeless hollow-core fiber (NANF) with a record low loss of 0.51 dB/km at 1.064 μm [95] and Jasion et al. reported the minimum loss of 0.28 dB/km between 1.51 and 1.60 μm in the NANF [96]. The HC-PCFs can provide a large light-gas interaction overlap inside hollow-core [6, 42] and wide transmission bands [39], exhibiting the potential for improving gas detection performance [101] and realizing a high-performance gas fiber laser [102-109].

On the other hand, the use of high-power lasers in the industry has been rapidly advanced by incorporating flexible-fiber-based beam delivery. The laser beams cannot be transported through traditional step-index fibers as their peak intensity lies above the damage threshold of silica and would therefore destroy the fiber itself. The HC-PCFs are now implemented as a tool of laser beam delivery in place of traditional optical fibers [110-116].

One of the problems arising from these high-power levels injected in the optical fiber is the probability of detonating the fiber fuse effect. This phonomenon was discovered by British researchers in 1987-1988 [117-121] in conventional single-mode optical fibers. The fiber fuse effect in kagome-lattice HC-PCFs [113] and revolver fibers [122, 123] has been reported.

In this paper, the generation of a fiber fuse and its propagation in kagome-lattice HC-PCFs and revolver fibers were investigated by the explicit finite-difference method using the thermochemical SiOx production model [124].

Kagome-Lattice HC-PCF

The occurrence of the fiber fuse in a kagome-lattice HC-PCF was first reported in 2014 by Dumitrache et al. [113]. They conducted high-power laser pulse delivery experiments using kagome-lattice HC-PCFs with a core diameter of 50 μm [62].

A schematic illustration of the fiber is shown in Figure 1.

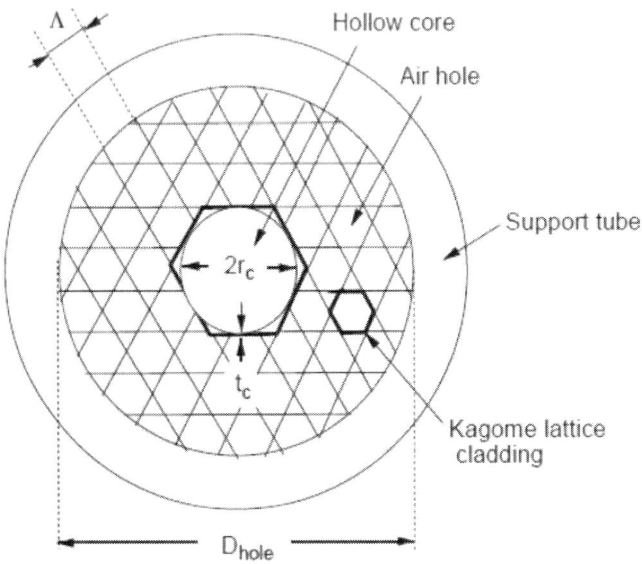

Figure 1. Schematic illustration of kagome-lattice HC-PCF.

In this figure, r_c (=25 μm) and D_{hole} (~164 μm) are the radius of the hollow core and the outer diameter of the kagome lattice cladding, respectively, and t_c (=0.28 μm) is the thickness of a silica strut of the kagome lattice. The fiber was first flushed with neon to remove any trapped dust inside the fiber (hollow) core [110].

Surface Damage at Fiber Tip

Dumitrache et al. were able to transmit as much as 30 mJ of energy through the fiber for 30 ns pulses at 1.064 μm [113]. The focused diameter of the fiber input was 41 μm.

However, damage occurred on the fiber tip when the input energy was increased to 35 mJ, and it propagated inward at a rate of about 0.0003 m/s when the laser was not immediately turned off. Laser-induced breakdown can be identified by the formation of an ionized gas or plasma during or by the end of the laser pulse [125]. When employing high-power pulses, the breakdown threshold intensity I_B for air at the focal minimum is determined by [125]

$$I_B = \frac{1}{\pi}\left(\frac{W_B}{r_0^2 \tau_p}\right), \qquad (1)$$

where W_B is the minimum laser pulse energy at which the breakdown is observed, r_0 is the focal spot radius, and τ_p is the duration or the full width at half-maximum (FWHM) of the laser pulse.

The I_B for air was estimated to be 8.8×10^{10} W/cm^2 when using $W_B = 35$ mJ, $r_0 = 20.5$ μm, and $\tau_p = 30$ ns. This value is close to the experimentally determined I_B (8.2×10^{10} W/cm^2 [126]). When the input energy of the laser radiation in the focal region becomes greater than the excitation energy for air, it will trigger breakdown of air molecules and cause ionization. In this initial breakdown (multiphoton ionization) process, a neutral atom absorbs enough laser photons to raise it from the ground state to the ionization level or above. Following the multiphoton ionization (MPI) process, the cascade effect takes place owing to the increased plasma density and the medium becomes opaque to the laser beam, leading to further energy absorption. This cascade ionization (CI) process produces a very sudden temperature rise in the medium at the focal region and results in the complete destruction of the internal structure of the fiber tip, as observed by Dumitrache et al.

Initiation Mechanisms of Fiber Fuse

On the other hand, fiber breakdown (fiber fuse) occurred at much lower energies (3.3 mJ and 7.6 mJ) than W_B (= 12 mJ) when employing a pulse duration of 12 ns [113].

This type of damage was very destructive. Once the fiber failed at the weak point, the damage propagated at a rate of about 0.05 m/s, also vaporizing, on its way, the kagome-lattice internal structure of the fiber.

Dumitrache et al. identified several mechanisms that may be responsible for this type of damage: laser power fluctuations and beam wandering, optical feedback caused by back-reflected light downstream of the fiber, and imperfections/microfractures inside the fiber [113].

Optical Feedback of Back-Reflected Light

Among them, optical feedback caused by back-reflected light will be fairly inhibited in the kagome-lattice HC-PCF.

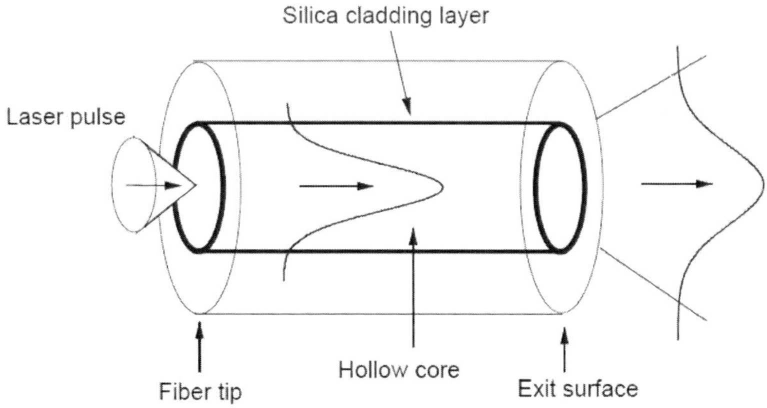

Figure 2. Schematic illustration of laser pulse propagation through kagome-lattice HC-PCF.

They used a seven-cell hypocycloid-core and three-ring kagome lattice cladding HC-PCF in their pulse delivery experiments. The air-filling fraction of the cladding structure was about 83% [49]. As can be seen from Figure 1, the occupancy of silica (about 15%) in the cross section of the core and cladding layers is small and the index of refraction n inside the support tube of the fiber is estimated to be about 1.0675, which is close to that (~1) of air at room temperature. By using n, we obtain the coefficient of reflection R at the exit surface (see Figure 2) of the fiber as

$$\mathrm{R} = \left(\frac{n-1}{n+1}\right)^2 \tag{2}$$

The R at the exit surface was estimated to be only about 0.001. This means that the quantity of back-reflected light along the longitudinal direction of the kagome-lattice HC-PCF is very small. Therefore, optical feedback caused by back-reflected light downstream of the fiber, which is one of the damage mechanisms proposed by Dumitrache et al. is considerably difficult to be realized in the kagome-lattice HC-PCF.

Physical Imperfections on Silica Strut Surface

It is a commonly observed phenomenon that surfaces tend to have a lower threshold for damage than the bulk material [127-131]. The ideas concerning the importance of the electric field strength at the surface of a dielectric material had been extended by Bloembergen to investigate the role of submicroscopic cracks, pores, and absorbing inclusions in the generation of surface damage [132, 133].

Physical imperfection on the surface of the silica strut is another of the damage mechanisms proposed by Dumitrache et al. First, we assume that there is a crack on the surface of the silica strut. If the crack is mathematically represented by a disk-shaped ellipsoidal void, with the lines of force in the silica strut normal to the disk, the electric field will be concentrated near the edge of the crack by a factor of n_s^2, where n_s (=1.45) is the refractive index of the silica [133]. That is, the laser pulse intensity I_c near the edge of the crack is given by

$$I_c = n_s^4 I_s \tag{3}$$

where I_s is the laser pulse intensity near the silica strut of the kagome-lattice cladding. I_s can be calculated by using

$$I_s = \eta \frac{1}{\pi}\left(\frac{W}{r_c^2 \tau_p}\right) \tag{4}$$

where W is the laser pulse energy and η is the theoretical optical power overlap ratio between the core and the cladding. Because the extinction ratio between the light intensity in the hollow (air) core and the kagome-lattice (silica) cladding is more than 40 dB [110], η becomes less than ~0.01%.

As shown in Eq. (3), the intensity of the light pulse near the edge of the crack on the surface is higher by a light intensity enhancement factor (LIEF) of n_s^4 (=5.06) than that near the silica strut. Furthermore, Génin et al. numerically calculated LIEF values for conical surface cracks [134]. They found that the LIEF values can locally reach two orders of magnitude higher than those for conical cracks of ideal shape and that theoretical LIEF values for the conical cracks range from 5 to 102 [134].

If we assume the maximum LIEF value (=102), I_c becomes 1.4×10^8 W/cm^2 using Eqs. (3) and (4), $W = 3.3$ mJ, $\tau_p = 12$ ns, and $\eta = 0.01\%$. This I_c value is three orders of magnitude smaller than the damage threshold intensity of silica [135]. This means that imperfections/microfractures inside the fiber, which is one of damage mechanisms proposed by Dumitrache et al. is not responsible for fiber fuse initiation in the kagome-lattice HC-PCF.

Absorbing Inclusions in Silica Strut

Next, we assume that there are absorbing inclusions in the surface of the silica strut. Commercial vitreous silica and soda-lime-silicate glasses contain metallic impurities such as iron and/or copper [136, 137]. The inclusion damage is associated with the presence of metallic absorbing particles in the silica glass.

When spherical metallic particles are considered, they are generally considered to be opaque at the laser frequency and absorb energy from the laser beam. If the laser pulse of the beam is taken to be a square wave of a given duration, the average temperature $\tilde{T}$ of the particle is expressed, neglecting any changes in phase, as [138]

$$\tilde{T} = \left(\frac{3\varepsilon I_S}{4\rho C_v R_i}\right) \cdot t, \tag{5}$$

where ε (= 0.3) is the spectral emissivity, and ρ, C_v, and R_i are the density, specific heat, and radius of the particle, respectively. In the case of Fe inclusions with the pulse energy W of 3.3 mJ and $t = \tau_p$ (= 12 ns), $\tilde{T}$ becomes

$$\tilde{T} \simeq 5.436 \times 10^{-5} / R_i, \tag{6}$$

where we used $\rho = 7{,}870$ kg m^{-3} and $C_v = 442$ J kg^{-1} K^{-1} [139]. As the thickness t_c of the silica strut is 0.28 μm, R_i must be smaller than t_c. If we assume R_i = 0.01 μm, $\tilde{T}$ becomes 5,436 K for a certain range of particle sizes.

As the silica glass is heated from room temperature, it melts in the vicinity of 1,996 K, turns into a liquid, and thermally decomposes near 3,000 K into a gas (SiO gas + O_2 gas) [140]. The estimated $\tilde{T}$ (=5,436 K) is higher than the decomposition temperature (about 3,000 K) of the silica.

Heat generated at the Fe particles is transmitted to the adjacent part of the silica strut, and the strut is heated. When the temperature of part of the silica strut rises to 3,000 K or higher, the optical absorption coefficient becomes on the order of 10^4 m^{-1}. As a result, heat generation due to light absorption also occurs in the strut. In this manner, the inclusion of metallic impurities in the silica struts is responsible for fiber fuse initiation, and the fiber fails at the weak point, that is, where the kagome-lattice internal structure had been destroyed by the vaporization of the silica struts.

As the metal migration velocity in the silica glass is very low (47 μm/s [141]), the damage propagation velocity (about 0.05 m/s) will be attributable to the thermal wave propagation along the longitudinal direction of the fiber.

In the following section, we describe the unsteady-state thermal conduction process in kagome-lattice HC-PCFs theoretically by the explicit finite-difference method on the basis of the thermochemical SiO_x production model [124].

Fiber Fuse Calculation of Kagome-Lattice HC-PCF

We used the simplified model shown in Figure 3 for heat conduction analysis of the kagome-lattice HC-PCF.

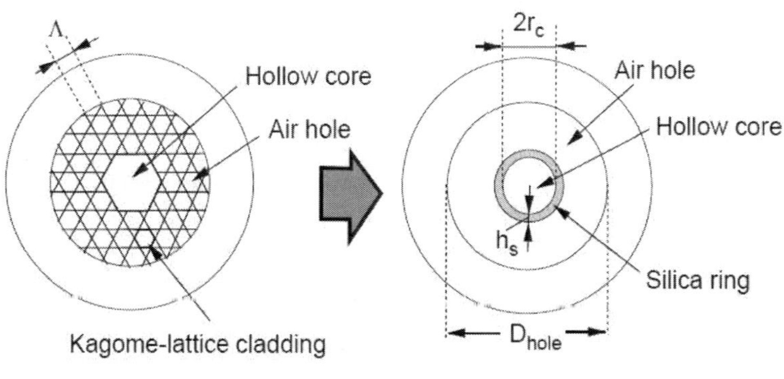

Figure 3. Heat conduction model for kagome-lattice HC-PCF.

The hollow core was assumed to be a cylindrical rod of diameter $2r_c$ and was surrounded by a suitably designed cladding of high (silica ring) and low (air hole) refractive indices. The hollow core and air hole layer were assumed to be filled with air at the same temperature as the surrounding air (T_a).

In an actual kagome-lattice HC-PCF, many struts of silica glass exist in the kagome-lattice cladding, and the air-filling fraction η_s of the cladding structure is about 83%. If the width h_s of the silica ring is set to 8 μm (~D_{hole}/20), η_s becomes about 92%. This η_s is slightly larger than that (83%) of an actual kagome-lattice HC-PCF.

We assumed that the kagome-lattice HC-PCF is in an atmosphere with temperature $T = T_a$ and part of the silica ring of length ΔL (=40 μm) is heated to a temperature of T_c^0 (>T_a) owing to light absorption of metallic impurities, as shown in Figure 4.

Thus, as light propagates along the positive direction (away from the light source) in this part, a considerable amount of heat is produced by light absorption.

The calculation by the finite-difference method followed the procedure described in the literature [124]. However, in the case of a kagome-lattice HC-PCF, since the air layer exists outside the silica-ring layer, it is necessary to consider the boundary conditions, as described below. When a certain quantity of heat per unit area (heat flux) is conducted from the heated silica-ring center to the end of the silica-ring layer ($r = r_c + h_s$) of an HC-PCF, a considerable amount of heat flux is transmitted through the air-hole layer with a temperature of T_a, and the residual heat flux stagnates in the core layer because the thermal conductivity κ of the air-hole layer is lower than that of silica glass (κ for air is about 1/300 that for glass). The stagnated heat flux in the silica-ring layer is used to heat air molecules existing in the hollow-core layer.

The heat transfer is a result of the air convection, which is generated by warming the air near the outer surface of the silica-ring layer. In this heat transfer, there is a region (called the "thermal boundary layer") near the outer surface of the silica-ring layer where the temperature of the air changes rapidly from a high value to T_a.

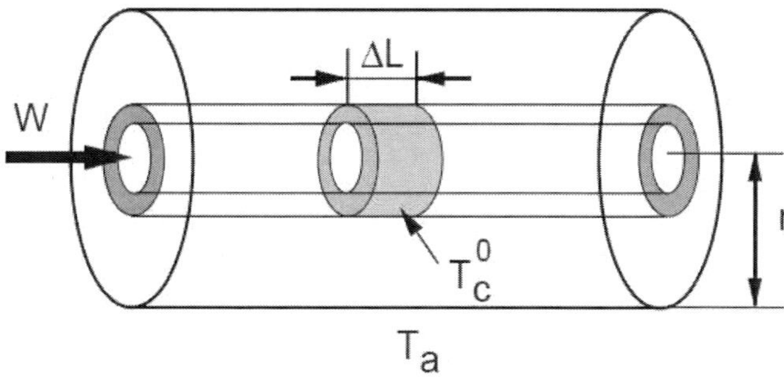

Figure 4. Hot zone in silica ring of kagome-lattice HC-PCF.

The boundary condition in this case is given by

$$-k\frac{\partial T}{\partial r}\bigg|_{r=r_c+h_s} = \sigma_s\varepsilon(T^4 - T_a^4) + \frac{k}{\delta r_t}(T - T_a) \quad (7)$$

where σ_s is the Stefan-Boltzmann constant and ε (~0.9) is the emissivity of the surface. δr_t is the thickness of the thermal boundary layer and $\delta r_t = x\ \delta r$ is assumed in the HC-PCF calculation, where δr is the step size along the r axis and $x = 0.5$.

In the calculation, we set the time interval δt to 10 ns, the step size along the r axis δr to $r_f/20$, and the step size along the z axis δz to 20 μm, and assumed that $r_f = D_{hole}/2$, $T_c^0 = 3{,}000$ K, and $T_a = 298$ K. We used the following values of κ (W m^{-1} K^{-1}), ρ (kg m^{-3}), and C_p (J kg^{-1} K^{-1}) of air in the heat conduction calculation [139].

$\kappa = 0.105$
$\rho = 0.183$
$C_p = 1{,}306$

The laser light intensity I at the silica ring was increased to 1 MW/cm^2 and higher; then the thermal wave increased in size and propagated in the negative z direction toward the light source. We estimated the temperature field T (r, z) of the kagome lattice HC-PCF at t = 1 ms and 6 ms after the incidence of laser light with I = 1.61 MW/cm^2 (=1.15 I_s) and λ_0 = 1.064 μm. The calculated temperature fields are shown in Figures 5 and 6.

As shown in Figure 5, the core center temperature near the end of the hot zone (z ~ -0.06 mm) increases abruptly to about 1.9 × 10^4 K after 1 ms. This rapid rise in the temperature initiates the fiber fuse propagation, as shown in Figure 6. After 6 ms, the high-temperature front in the core reaches a *z* value of -0.58 mm. The average propagation velocity V_f was estimated to be 0.10 m/s using these data. This V_f is two times larger than that (0.05 m/s) observed by Dumitrache et al. [113] and it is about one-tenth that (~1 m/s) of the SMF28 optical fiber [142].

If there are absorbing inclusions such as Fe particles in the surface of the silica ring, heat generated at the Fe particles is transmitted to the adjacent part of the silica ring, and the temperature of that part of the silica ring rises to 3,000 K. When *I* at the heated silica ring is increased to 1 MW/cm^2 and above, the thermal wave increases in size and propagates in the negative *z* direction toward the light source. In this way, the damage, starting at the heated part of the silica ring, propagates through the silica ring along the longitudinal direction of the fiber.

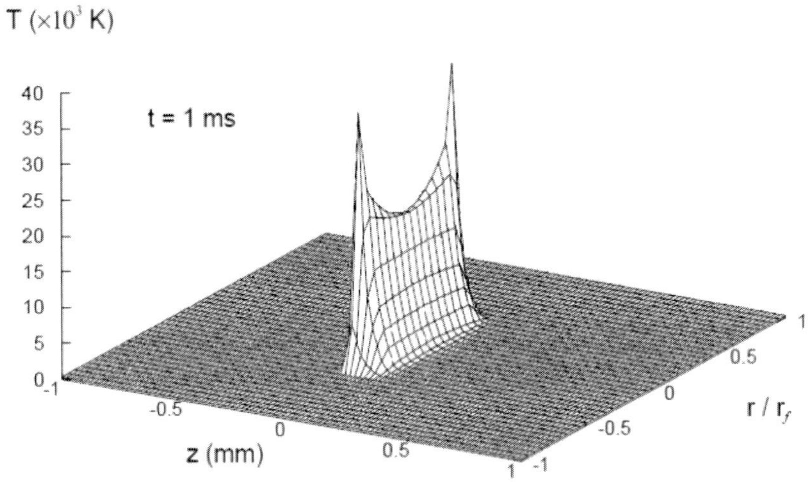

Figure 5. Temperature field in kagome-lattice HC-PCF after 1 ms when I = 1.61 MW/cm^2 at λ_0 = 1.064 μm.

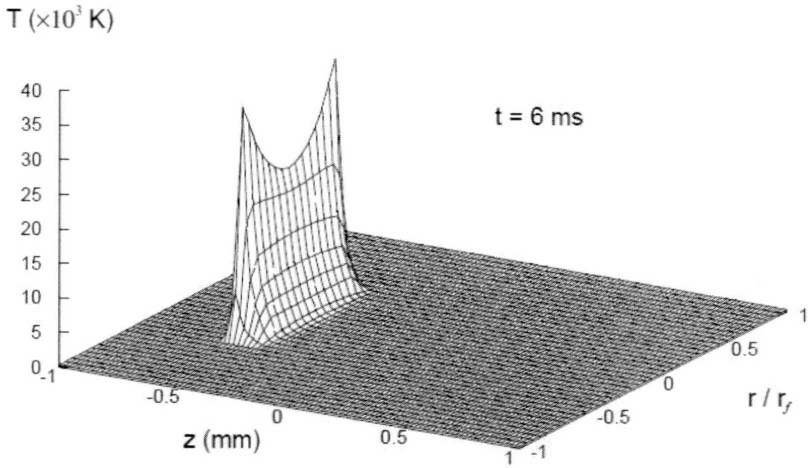

Figure 6. Temperature field in kagome-lattice HC-PCF after 6 ms when I = 1.61 MW/cm^2 at λ_0 = 1.064 μm.

Revolver Fibers

The occurrence of a fiber fuse in a revolver fiber was first reported in 2018 by Kolyadin et al. [122]. They fabricated the two types of revolver fibers (RF1 and RF2) shown in Figure 7.

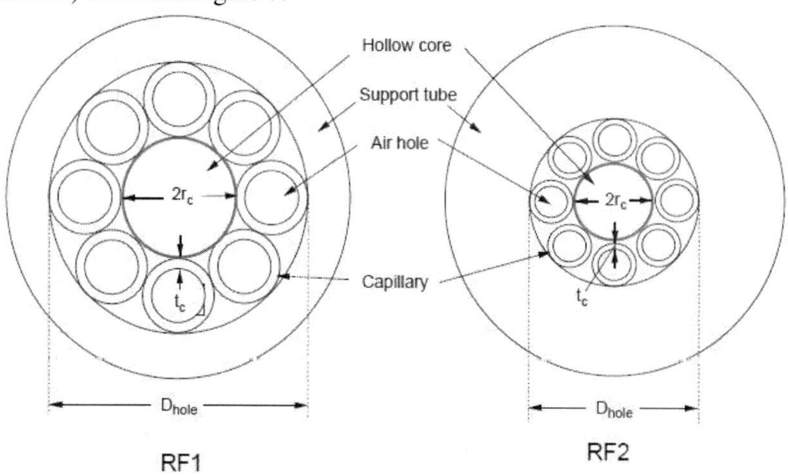

Figure 7. Two types of revolver fiber.

Both fibers have a relatively simple air-silica cladding structure, which consists of one layer of capillaries. In RF1, these capillaries touch each other, whereas they are located at a certain distance in RF2. In Figure 7, the hollow core radius r_c, the inner diameter of the support tube D_{hole}, and the capillary thickness t_c of RF1/RF2 are 42/20 μm, 93/36 μm, and 3.1/0.8 μm, respectively. The support tube diameter ($2r_f$) of RF1/RF2 is 125/100 μm.

Laser Source for Fiber Fuse Experiments

The fiber fuse experiments were carried out using a pulse laser operated at λ_0 = 1.064 μm. The laser radiation consisted of nanosecond trains of picosecond pulses (NTPPs) and had the duration of 130 ns. The fiber fuse initiation was observed in RF1 and RF2 at the average powers of about 4 W (RF1) and about 2 W (RF2), and the fiber fuse propagated at a velocity of about 1 m/s. The threshold power corresponded to the average intensity of 2.4×10^9 W/cm^2 (RF1) and 5.2×10^9 W/cm^2 (RF2) over the duration of the NTPP. These values were one order of magnitude smaller than the breakdown threshold intensity I_B (8.2×10^{10} W/cm^2 [126]) for air, which was estimated for the 10 ns pulsed laser operated at λ_0 = 1.064 μm.

For a revolver fiber, it is known that the extinction ratio of the power in the center of the hollow core to that in the silica capillary is more than 20 dB [75]. This means that the theoretical optical power overlap ratio between the core and the cladding, η, becomes less than 1%. If η of 0.5% is assumed for the revolver fiber, the intensities I_s near the silica capillaries in RF1 and RF2 can be estimated to be 12 MW/cm^2 (RF1) and 26 MW/cm^2 (RF2) over the duration of the NTPP. These values are of the same magnitude as the threshold intensity I_{th} (about 1-10 MW/cm^2) observed in the fiber fuse experiments using a CW laser operated at λ_0 = 1.064 μm [142].

Fiber Fuse Initiation in Revolver Fiber

In the fiber fuse experiments, a planar absorbing metal (absorbent) surface was brought parallel to the fiber end face [122]. In what follows, we used copper as the absorbing metal. When the laser beam was focused on the absorbent, the temperature of the laser-illuminated spot on the absorbent rose. The optical absorption coefficient α is related to the extinction coefficient k and λ_0 by [143]

$$\alpha = 4\pi\, k/\lambda_0 \qquad (8)$$

A value of $k = 6.22$ for copper at $\lambda_0 = 0.96$ μm has been reported [144]. Using Eq. (8) and this k value, the absorption coefficient of copper was estimated to be $\alpha = 7.35 \times 10^7$ m^{-1} at $\lambda_0 = 1.064$ μm. This value is about three orders of magnitude larger than the α values (10^4 m^{-1} order) required for fiber fuse generation.

The part of the silica capillaries in contact with the heated absorbent is also heated by thermal conduction and radiation from the absorbent, as described in [145]. Then, this part starts to absorb the laser beam because of its absorption enhancement due to the temperature rise. The heated part of the capillaries also heats the surrounding area, which also starts to absorb the laser beam. As this process is repeated, the heated spot will move backward (toward the light source), as observed by Kolyadin et al. [122].

In the following section, we describe the unsteady-state thermal conduction process in a revolver fiber theoretically by the explicit finite-difference method on the basis of the thermochemical SiO_x production model [124].

Fiber Fuse Calculation in Revolver Fibers

We used the simplified model shown in Figure 8 for heat conduction analysis of the revolver fiber. This model is similar to the multi-layered model proposed by Wang and Ding [100]. However, our model simplified the structure of revolver fiber, compared with their multi-layered model. The hollow core was assumed to be a cylindrical rod of diameter $2r_c$ and it was surrounded by a suitably designed cladding of high (silica ring) and low (air hole) refractive indices. The hollow core and air hole layer were assumed to be filled with air at the same temperature as the surrounding air (T_a).

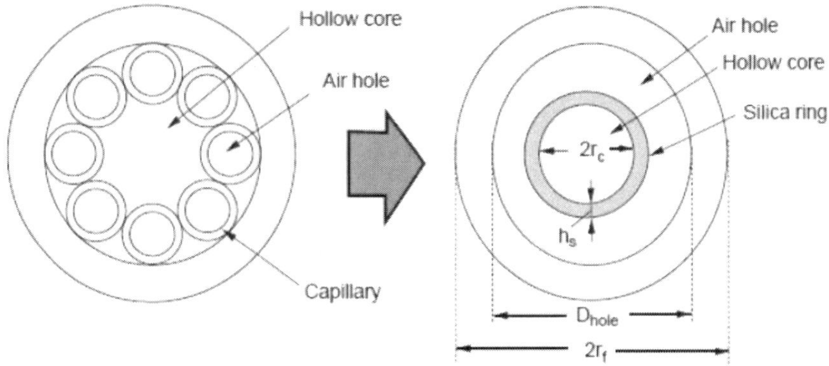

Figure 8. Heat conduction model for revolver fiber RF1.

Fiber Fuse Propagation in RF1

In an actual revolver fiber RF1, eight capillaries of silica glass exist in the air-silica cladding, and the air-filling fraction η_s of the cladding structure is about 68%. If the width h_s of the silica ring is set to 11 μm (~$D_{hole}/8$), η_s becomes about 66%. This η_s is slightly smaller than that (68%) of an actual RF1.

Using the model described above, we investigated the appearance of the fiber fuse effect in a revolver fiber by the explicit finite-difference method. We assumed that the revolver fiber is in an atmosphere of temperature $T = T_a$ and part of the silica ring of length ΔL (=40 μm) is heated to a temperature of T_c^0 (>T_a), as shown in Figure 4.

In the calculation, we set the time interval δt to 10 ns, the step size along the r axis δr to $r_f/20$, and the step size along the z axis δz to 20 μm and assumed that T_c^0 = 3,000 K and T_a = 298 K.

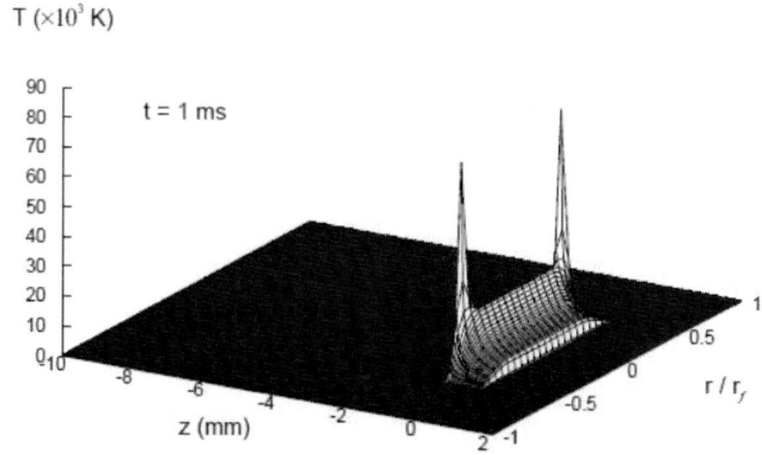

Figure 9. Temperature field in RF1 after 1 ms when I = 18 MW/cm^2 at λ_0 = 1.064 μm.

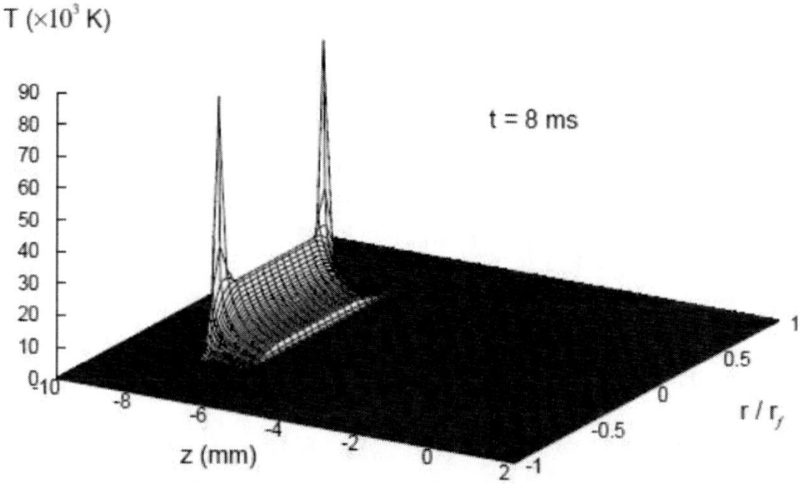

Figure 10. Temperature field in RF1 after 8 ms when I = 18 MW/cm^2 at λ_0 = 1.064 μm.

We estimated the temperature field T (r, z) of the revolver fiber RF1 at t = 1 ms and 8 ms after the incidence of laser light with I = 18 MW/cm^2 (=1.5 I_s) and λ_0 = 1.064 μm. The calculated temperature fields are shown in Figures 9 and 10.

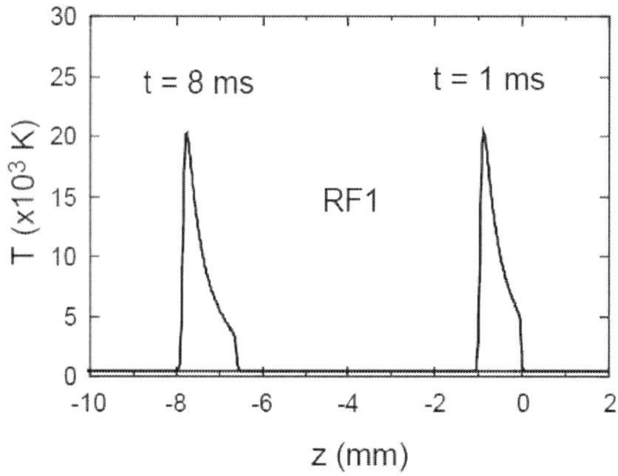

Figure 11. Temperature fields of core center of RF1 after 1 ms and 8 ms when I = 18 MW/cm^2 and λ_0 = 1.064 μm.

As shown in Figure 9, the temperature of the silica-ring center near the end of the hot zone (z = -0.96 mm) increases abruptly to about 7.1 × 10^4 K after 1 ms. The heated part of the silica ring heats air molecules in the hollow core, in which the temperature also starts to increase. This rapid rise in the temperature initiates the fiber fuse propagation, as shown in Figure 10. After 8 ms, the high-temperature front in the silica-ring center reaches a z value of -7.84 mm. The average propagation velocity V_f was estimated to be 0.98 m/s using these data. This value is close to the V_f (1.0 m/s) observed by Kolyadin et al. [122] and is also the same value as that (~1 m/s) of the SMF28 optical fiber [142].

The temperature fields of the core center along the z direction were calculated at t = 1 ms and 8 ms after the incidence of laser light with I = 18 MW/cm^2 and λ_0 = 1.064 μm. The calculated results are shown in Figure 11.

The heat generated in the silica-ring layer is transferred to the neighboring core layers of the RF1. At a time of 1 ms after laser light incidence, a peak temperature (T_p) of 2.0 × 10^4 K or above occurs in the center of the hollow-core layer. The T_p value is maintained after 8 ms, as shown in Figure 11. The propagation velocity of the heated air is the same as that (0.98 m/s) of fiber fuse propagation.

Fiber Fuse Propagation in Polymer-Coated RF2

Next, we investigated the unsteady-state thermal conduction process in the revolver fiber RF2 theoretically by the explicit finite-difference method. We used the simplified model shown in Figure 8 for heat conduction analysis of the revolver fiber RF2.

In an actual revolver fiber RF2, eight capillaries of silica glass exist in the air-silica cladding, and the air-filling fraction η_s of the cladding structure is about 79%. If the width h_s of the silica ring is set to 5 μm (~$D_{hole}/7$), η_s becomes about 44%. This η_s is smaller than that (79%) of an actual RF2.

We assumed that the revolver fiber RF2 is in an atmosphere of temperature $T = T_a$, and part of the silica ring of length ΔL (=40 μm) is heated to a temperature of T_c^0 ($>T_a$), as shown in Figure 4. In the calculation, we set the time interval δt to 5 ns, the step size along the r axis δr to $r_f/20$, and the step size along the z axis δz to 20 μm, and assumed that T_c^0 = 3,000 K and T_a = 298 K.

We estimated the temperature fields of the core center of the revolver fiber RF2 along the z direction at t = 1 ms and 8 ms after the incidence of laser light with I = 24 MW/cm^2 (= 0.93 I_s) and λ_0 = 1.064 μm. The calculated results are shown in Figure 12. As shown in Figure 12, the core center temperature near the end of the hot zone (z = -0.86 mm) increases abruptly to about 5.2 × 10^4 K after 1 ms. This rapid rise in the temperature initiates the fiber fuse propagation. After 8 ms, the high-temperature front in the core center reaches a z value of -7.14 mm. The average propagation velocity V_f was estimated to be 0.90 m/s using these data. This value is close to the V_f (0.91 m/s) observed by Kolyadin et al. [122]. In RF2, the T_p value (5.2 × 10^4 K) is maintained after 8 ms, as shown in Figure 12.

The temperature field near the high-temperature front after the incidence of laser light with I = 24 MW/cm^2 was calculated along the r direction at t = 1 ms. The calculated result is shown in Figure 13.

As shown in Figure 13, the temperatures in the hollow core are higher than 3 × 10^4 K, and the temperatures in the silica-ring layer decrease with increasing r and reach ~T_a at the end of the silica-ring layer. The temperature distribution shown in Figure 13 is similar to the intensity distribution of optical-discharge plasma radiation in RF2 (Figure 5 (b) in [122]) observed by Kolyadin et al.

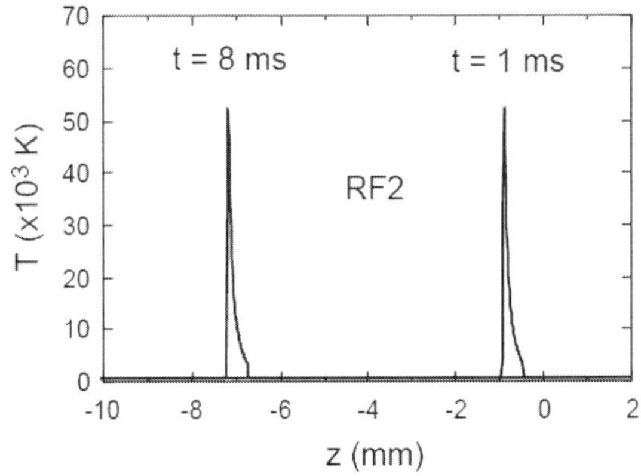

Figure 12. Temperature fields of core center of RF2 after 1 ms and 8 ms when I = 24 MW/cm^2 and λ_0 = 1.064 μm.

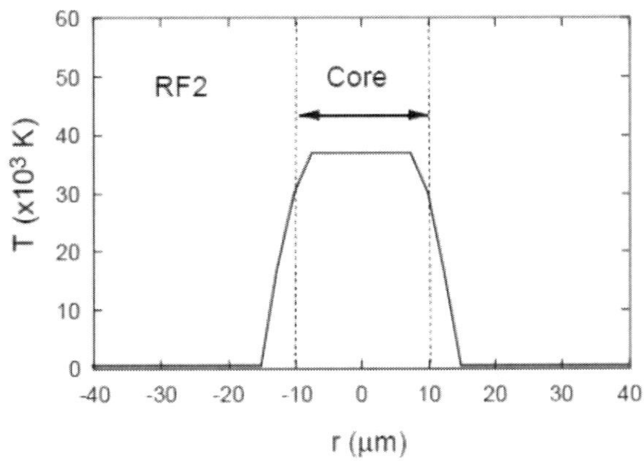

Figure 13. Temperature field near high-temperature front of RF2 after 1 ms when I = 24 MW/cm^2 and λ_0 = 1.064 μm.

Fiber Fuse Propagation in Uncoated RF2

If RF2 was completely covered with a polymer coating, the measured V_f was 0.91 m/s [122]. However, if the polymer coating was partially removed and the silica support tube directly bordered air, the measured V_f became 2.84 m/s, which was three times larger than that of the polymer-coated RF2 [123].

RF2 consists of a hollow core and an air-silica cladding, which contains one layer of silica capillaries (see Figure 7). In this fiber, the transverse components of a light wave along the radial direction propagate from the core to the silica capillary, and are then sent back to the core after being reflected at the core-capillary boundary and the inner and outer surfaces of the silica capillaries, as shown in Figure 14 (a),

If the polymer coating of the RF2 is removed, the Fresnel reflection at the outer surface of the support tube occurs and the back-reflected light wave is incident upon the silica capillaries and the hollow core, as shown in Figure 14 (b).

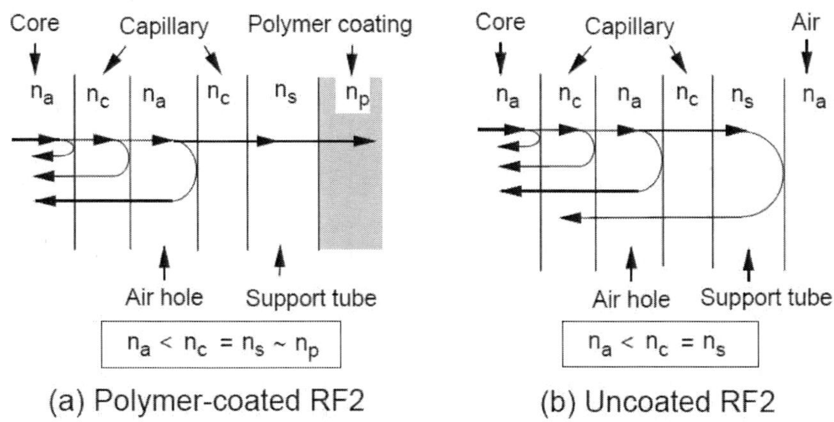

Figure 14. Schematic of light localization in core of polymer-coated RF2 (a) and uncoated RF2 (b).

If all or several of the reflected waves stated above are in phase at the core-capillary boundary, then they are mutually enhanced as a consequence of constructive interference.

To clarify the in-phase condition, we estimated four phase changes Φ_0-Φ_3 of optical routes shown in Figure 15. Among them, Φ_3 is solely related to the uncoated RF2.

The phase change is accumulated by the radial translation and an additional phase change Φ_R due to the reflection [146]. Φ_R depends on the relative magnitude in the refractive indices of layers situated at both sides of the interface. If a light wave travels from a low-index material to a high-index material, Φ_R becomes 0. Therefore, Φ_0 in Figure 15 leads to 0. On the contrary,

Φ_R becomes π when a light wave travels from a high-index material to a low-index material.

To determine the phase change due to radial translation, we need the lateral propagation constants denoted by

$$k_i = 2\pi\sqrt{n_i^2 - n_{eff}^2} / \lambda_0 \; (i = a, c, s) \tag{9}$$

where n_{eff} is the effective index of the least lossy mode of RF2 and λ_0 (= 1.064 μm) is the wavelength of the light wave.

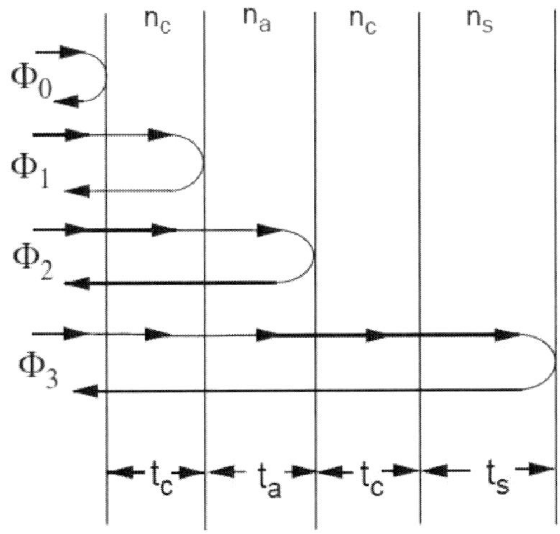

Figure 15. Optical routes in core of polymer-coated or uncoated RF2.

As n_{eff} of the fundamental mode is close to n_a (= 1.0) [147, 148] at λ_0 = 1.064 μm, Eq. (9) becomes

$$k_i \approx 2\pi\sqrt{n_i^2 - 1} / \lambda_0 \; (i = a, c, s) \tag{10}$$

The refractive index n of pure silica can be calculated using the Sellmeier equation:

$$n^2 - 1 = \sum_{i=1}^{3} c_i \, \lambda_0^2 / (\lambda_0^2 - b_i) \tag{11}$$

where c_i and b_i are the parameters listed in Table 1 [149].

The n value of pure silica at $\lambda_0 = 1.064$ μm was calculated to be 1.44988 using Eq. (11). Therefore, in the calculation, we used $n_c = n_s = 1.452$ for the silica capillary and silica support tube at $\lambda_0 = 1.064$ μm. We also used $n_a = 1.0$, $t_c = 0.8$ μm, $t_a = 6.4$ μm, and $t_s = 32.0$ μm.

Table 1. Sellmeier parameters of pure silica [149]

Parameters	Pure SiO_2
c_1	0.6965325
c_2	0.4083099
c_3	0.8968766
b_1	0.004368309
b_2	0.01394999
b_3	97.93399

Let us consider the phase change Φ_1-Φ_3 shown in Figure 15. To satisfy the in-phase condition responsible for Φ_0, waves reflected from the outer and inner surfaces of the silica capillary and the inner surface of the support tube must satisfy the following relations (Bragg conditions):

$$\Phi_1 - \Phi_0 = 2\, k_c\, t_c + \pi = 2\, \pi\, m_1 \tag{12}$$

$$\Phi_2 - \Phi_0 = 2\, (k_c\, t_c + k_a\, t_a) = 2\, \pi\, m_2 \tag{13}$$

$$\Phi_3 - \Phi_0 = 2\, (k_c\, t_c + k_a\, t_a + k_c\, t_c + k_s\, t_s) + \pi = 2\, \pi\, m_3 \tag{14}$$

where m_1, m_2, and m_3 are integers. By substituting Eq. (10) into Eqs. (12)-(14), the following equations are derived:

$$nt_1 \equiv 4\sqrt{n_c^2 - 1}\, t_c = (2\, m_1 - 1)\, \lambda_0 \tag{15}$$

$$nt_2 \equiv 2\sqrt{n_c^2 - 1}\, t_c = m_2\, \lambda_0 \tag{16}$$

$$nt_3 \equiv 4\, (2\sqrt{n_c^2 - 1}\, t_c + \sqrt{n_s^2 - 1}\, t_s) = (2\, m_3 - 1)\, \lambda_0 \tag{17}$$

We calculated nt_1-nt_3 and estimated m_1-m_3 using Eqs. (15)-(17). The calculated results are shown in Table 2.

As shown in this table, m_2 (=1.58) is not an integer, but m_1 (=2.08) and m_3 (=66.99) can be regarded as the integers 2 and 67.

This means that the reflected waves from the outer surfaces of the silica capillary and support tube are in phase at the core-capillary boundary, and they are mutually enhanced as a consequence of the constructive interference.

As a result, the power in the hollow core and the silica capillary will be improved for the uncoated RF2 compared with the polymer-coated RF2. If the gain in power of 5 dB is assumed for the uncoated RF2, the intensity I_s near the silica capillaries in the uncoated RF2 can be estimated to be 76 MW/cm^2 over the duration of the NTPP.

We estimated the temperature fields of the core center of the uncoated RF2 along the z direction at t = 1 ms and 4 ms after the incidence of laser light with I = 76 MW/cm^2 and λ_0 = 1.064 μm. The calculated results are shown in Figure 16.

Table 2. Calculated parameters of phase conditions

Parameter	Unit	Value
nt_1	μm	3.369
nt_2	μm	1.684
nt_3	μm	141.491
m_1	-	2.08
m_2	-	1.58
m_3	-	66.99

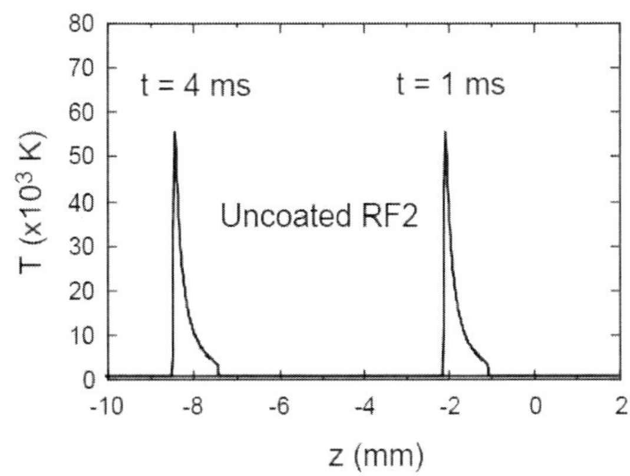

Figure 16. Temperature fields of core center of uncoated RF2 after 1 ms and 4 ms when I = 76 MW/cm^2 and λ_0 = 1.064 μm.

As shown in Figure 16, the core center temperature near the end of the hot zone (z = -2.06 mm) increases abruptly to about 5.5 × 10^4 K after 1 ms. This rapid rise in the temperature initiates the fiber fuse propagation. After 4 ms, the high-temperature front in the core center reaches a z value of -8.38 mm. The average propagation velocity V_f was estimated to be 2.11 m/s using these data. This value is close to the V_f (2.84 m/s) observed by Bufetov et al. [123]. In the uncoated RF2, the T_p value (5.5 × 10^4 K) is maintained after 4 ms, as shown in Figure 16.

Cavity Formation in Revolver Fiber

When a fiber fuse was generated in RF2, the damage was in the form of periodic (or nonperiodic) cavities remaining in the fiber, and the periodic cavity interval Λ of the damage was about 170 μm [122] or 180 μm [123]. The periodicity of the cavities was sometimes significantly disturbed [123].

This Λ is about one order larger than that (13-22 μm) [150, 151] observed in conventional single-mode optical fibers (SMFs) with a CW laser source.

When CW laser light is incident upon an SMF, heat generated continuously at the hot zone in the core layer owing to light absorption and the temperature of this part rises to 1 × 10^4 K and above. Then the thermal wave increases in size and starts to propagate in the negative z direction toward the light source. In this process, the periodic cavities with (relatively) small Λ values are formed in the SMF.

The relationship between the heat penetration time t_p and temperature penetration thickness δ in a silica glass layer is given by (see Appendix)

$$\delta{\sim}3.6\sqrt{kt_p}, \tag{18}$$

where k is the thermal diffusivity. k values of silica glass at 500 K and 3,000 K are 4.40 × 10^{-6} m^2/s and 8.32 × 10^{-8} m^2/s, respectively. The relationship between t_p and δ in a silica glass layer was investigated at 500 K and 3,000 K. The calculated results are shown in Figure 17.

As shown in this figure, t_p values increase with increasing δ and the t_p values at 3,000 K are larger than those at 500 K. The t_p value at δ = 60 μm is about 3.3 ms at 3,000 K.

In the fiber fuse experiments of the RF2, a Nd:YAG laser with a repetition rate of 1,200 Hz was used as a pulse laser source. This means that the period of the NTPP is about 0.83 ms. This period is shorter than the heat penetration time (3.3 ms) through a 60-μm-thick glass layer.

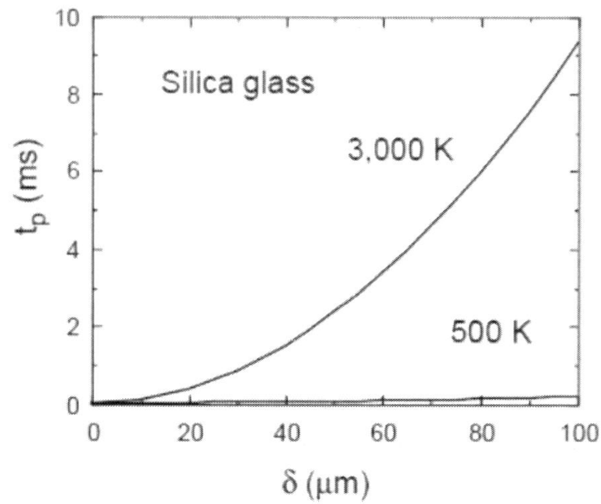

Figure 17. Relationship between tp and δ in silica glass layer.

On the other hand, it is well known that the first cavity length of fiber fuse propagation is a large value of 160-180 μm [151]. The duration of NTPPs is 130 ns [122, 123]. This duration is about one-tenth the time (18-30 μs) [151] for the formation of second (short) cavity from the first long cavity during fiber fuse propagation.

When an NTPP is incident upon the RF2, heat generated at the silica capillary owing to light absorption is transmitted to the adjacent part of the capillary, and the temperature of the part rises to 3,000 K. Then the thermal wave increases in size and starts to propagate in the negative z direction toward the light source. In this way, the first long cavity is formed in the RF2. Then the heat in the capillaries is efficiently transferred in the air of the adjacent core and the long cavity is frozen after a quick shutdown of the laser source until the next NTPP arrives at RF2 after 0.83 ms. In this way, the formation of a long cavity with a large Λ (about 170 μm or 180 μm) is allowed to continue for each incidence of the NTPP upon the RF2.

Bufetov et al. [123] explained the generation of a large Λ during fiber fuse propagation by adopting the concept of a laser-supported detonation wave (LSDW) and an attenuating shock wave (SW) as follows.

Under the action of a picosecond (ps) laser pulse with the maximum intensity of 7.0×10^{11} W/cm^2, gas plasma is produced in the air by bringing the output end of RF2 into contact with a metal target; then an LSDW starts to propagate along the fiber towards the light source. The LSDW continues its movement as an attenuating SW after the end of the ps pulse. The maximum intensity (7.0×10^{11} W/cm^2) of the ps pulse is one order of magnitude larger than the breakdown threshold intensity (8.2×10^{10} W/cm^2 [126]) for air. Therefore, the point explosion due to the optical breakdown of air will occur around major peaks of ps pulses. The motion of the gas plasma after the cessation of the laser pulse can be estimated on the basis of the point explosion theory with spherical symmetry [152, 153].

According to this theory, the gas plasma in the SW expands as

$$L(t) = \left(\frac{W}{\rho_0}\right)^{1/5} \cdot t^{2/5}, \tag{19}$$

where L is the distance from the center of the explosion, W (=0.17 mJ) is the energy of the ps pulse, and ρ_0 (=1.3 kg/m^3) is the initial density of air in the core of RF2. As the period of the ps pulses is 13 ns, L at t = 13 ns is estimated to be about 117 μm using Eq. (19). This L value is close to the Λ value (about 170 μm or 180 μm) observed in RF2.

When the next ps pulse arrives 13 ns later, the expanded gas plasma of the SW is not completely relaxed in the hollow core. Hence, the next pulse is absorbed in the front of a still-powerful SW. As a result, the propagation of the LSDW and its relaxation are repeated, forming several cavities with large Λ along the fiber towards the light source. The evolution number N_{pe} of the point explosion during the period (0.83 ms) of the NTPP is highly affectd by laser power fluctuation. The changes in N_{pe} result in a disturbance of the periodicity of the cavities, as observed by Bufetov et al. [123].

Conclusion

Hollow-core photonic crystal fibers (HC-PCFs) can guide light inside the hollow core by photonic bandgap or inhibited-coupling mechanisms. HC-PCFs are now implemented as an important tool of laser beam delivery in place of traditional single mode optical fibers. One of the problems arising from high-power injection in the optical fiber is the probability of triggering the fiber fuse effect. This effect is related to a heat conduction process in optical fibers. The unsteady-state thermal conduction process in several HC-PCFs was studied theoretically by the explicit finite-difference method using the thermochemical SiOx production model. For heat conduction analysis, the complicated inner structure of HC-PCFs was simplified using the model composed of silica-ring and air-hole layers. In a kagome-lattice HC-PCF, the inclusion of metallic impurities in the silica struts was responsible for fiber fuse initiation. The calculated velocities of fiber fuse propagation in the kagome-kattice PCF were in fair agreement with the experimental values. Furthermore, the calculated velocities of fiber fuse propagation in two types of polymer-coated revolver fibers were in fair agreement with the experimental values. If the polymer coating of the revolver fiber is removed, the Fresnel reflection at the outer surface of the support tube occurs and the back-reflected light wave is incident upon the silica capillaries and the hollow core. To clarify the in-phase condition, we estimated phase changes of optical routes in the uncoated revolver fiber. It was found that the reflected waves from the outer surfaces of the silica capillary and support tube are in phase at the core-capillary boundary, and they are mutually enhanced as a consequence of the constructive interference. As a result, the power in the hollow core and the silica capillary was improved for the uncoated revolver fiber compared with the polymer-coated one. The calculated velocity of fiber fuse propagation in the uncoated revolver fiber was in fair agreement with the experimental value.

Appendix: Temperature Penetration Thickness

The semi-infinite solid provides a useful idealization for many practical problems. It is used to approximate the transient response of a finite solid, such as a thin silica glass layer. The unsteady heat conduction equation in a semi-infinite solid is given by

$$\frac{\partial T}{\partial t} = \mathrm{k}\frac{\partial^2 T}{\partial x^2}, \mathrm{k} = \lambda / \rho\, C_{\mathrm{p}}. \tag{A1}$$

k is the thermal diffusivity and λ, ρ, and C_p are the thermal conductivity, density, and specific heat, respectively. To solve Eq. (A1) for the temperature distribution T (x, t), it is necessary to specify an initial dondition and two interior boundary conditions. The initial condition is

$$T(x, 0) = T_0 \quad \text{(A2)}$$

and the boundary conditions are

$$T(0, t) = T_s,\ T(\infty, t) = T_0 \quad \text{(A3)}$$

The solution is obtained by introducing a similarity variable $\eta = x / 2\sqrt{kt}$, through which the heat equation is transformed from a partial differential equation, involving two independent variables (x and t), to an ordinary differential equation expressed in terms of the single similarity variable [154].

Using η, Eq. (A1) becomes

$$\frac{\partial^2 T}{\partial \eta^2} + 2\eta \frac{\partial T}{\partial \eta} = 0\ . \quad \text{(A4)}$$

The general solution of Eq. (A4) is given by

$$T = C_1 \cdot erf(\eta) + C_2, \quad \text{(A5)}$$

where C_1 and C_2 are the undetermined constants and $erf(\eta)$ is the Gaussian error function, which is given by

$$erf(\eta) = \frac{2}{\sqrt{\pi}} \int_0^{\eta} \exp(-u^2)\, du \quad \text{(A6)}$$

C_1 and C_2 are determined by applying the intial and boundary conditions as

$$C_1 = T_0 - T_s,\ C_2 = T_s. \quad \text{(A7)}$$

By substituting Eq. (A7) into Eq. (A5), the following equation is derived

$$\frac{T - T_0}{T_s - T_0} = 1 - erf(\eta) \quad \text{(A8)}$$

Temperature distribution in a semi-infinite solid was estimated using Eq. (A8). The calculated result is shown in Figure A.1.

As shown in this figure, $(T - T_s) / (T_s - T_0)$ decreases with increasing η and gradually approaches zero at $\eta = 3$.

The temperature penetration thickness δ and heat penetration time t_p are defined as the distance x and time t at which $(T - T_s) / (T_s - T_0) = 0.01$ and $\eta \simeq 1.8$ [155]. When $\eta \simeq 1.8$, the relationship between δ and t_p is given by

$$\delta \sim 3.6\sqrt{kt_p}\,. \tag{A9}$$

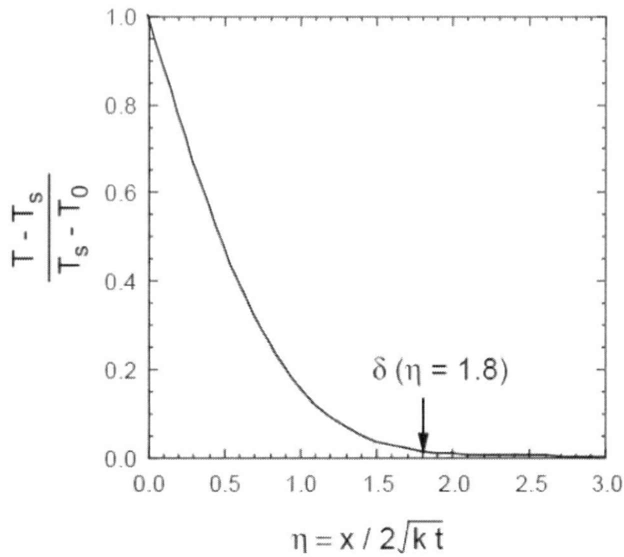

Figure A.1. Temperature distribution in semi-infinite solid.

References

[1] Birks T. A., Roberts P. J., Russell P. St. J., Atkin D. M., and Shepherd T. J. (1995). Full 2-D photonic bandgaps in silica/air structures. *Electron. Lett.*, 31 (22), 1941-1943.

[2] Knight J. C., Birks T. A., Russell P. St. J., and Atkin D. M. (1996). All-silica single-mode optical fiber with photonic crystal cladding. *Opt. Lett.*, 21 (19), 1547-1549.

[3] Birks T. A., Knight J. C., and Russell P. St. J. (1997). Endlessly single-mode photonic crystal fiber. *Opt. Lett.*, 22 (13), 961-963.

[4] Mogilevtsev D., Birks T. A., and Russell P. St. J. (1998). Group-velocity dispersion in photonic crystal fibers. *Opt. Lett.*, 23 (21), 1662-1664.

[5] Knight J. C., Birks T. A., Russell P. St. J., and de Sandro J. P. (1998). Properties of photonic crystal fiber and the effective index model. *J. Opt. Soc. Am. A*, 15 (3), 748-752.

[6] Knight J. C., Broeng J., Birks T. A., and Russell P. St. J. (1998). Photonic band gap guidance in optical fibers. *Science*, 282, 1476-1478.

[7] Cregan R. F., Mangan B. J., Knight J. C., Birks T. A., Russell P. St. J., Roberts P. J., and Allan D. C. (1999). Single-mode photonic band gap guidance of light in air. *Science*, 285, 1537-1539.

[8] Bennett P. J., Monro T. M., and Richardson D. J. (1999). Toward practical holey fiber technology: fabrication, splicing, modeling, and characterization. *Opt. Lett.*, 24 (17), 1203-1205.

[9] Broeng J., Mogilevstev D., Barkou S. E., and Bjarklev A. (1999). Photonic crystal fibers: a new class of optical waveguides. *Opt. Fiber Technol.*, 5, 305-330.

[10] Monro T. M., Richardson D. J., Broderick N. G. R., and Bennett P. J. (1999). Holey optical fibers: an efficient modal model. *IEEE J. Lightwave Technol.*, 17 (6), 1093-1102.

[11] Broderick N. G. R., Monro T. M., Bennett P. J., and Richardson D. J. (1999). Nonlinearity in holey optical fibers: measurement and future opportunities. *Opt. Lett.*, 24 (20), 1395-1397.

[12] Birks T. A., Mogilevtsev D., Knight J. C., and Russell P. St. J. (1999). Dispersion compensation using single-material fibers. *IEEE Photon. Technol. Lett.,* 11 (8), 674-676.

[13] Ranka J. K., Windeler R. S., and Stentz A. J. (2000). Visible continuum generation in air-silica microstructure optical fibers with anomalous dispersion at 800 nm. *Opt. Lett.*, 25 (1), 25-27.

[14] Broeng J., Barkou S. E., Sonbergaard T., and Bjarklev A. (2000). Analysis of air-guiding photonic bandgap fibers. *Opt. Lett.*, 25 (2), 96-98.

[15] Ranka J. K., Windeler R. S., and Stentz A. J. (2000). Optical properties of high-delta air-silica microstructure optical fibers. *Opt. Lett.,* 25 (11), 796-798.

[16] Wadsworth W. J., Knight J. C., Ortigosa-Blanch A., Arriaga J., Silvestre E., and Russell P. St. J. (2000). Soliton effects in photonic crystal fibres at 850 nm. *Electron. Lett.*, 36 (1), 53-55.

[17] Knight J. C., Arriaga J., Birks T. A., Ortigosa-Blanch A., Wadsworth W. J., and Russell P. St. J. (2000). Anomalous dispersion in photonic crystal fiber. *IEEE Photon. Technol. Lett.*, 12 (12), 807-809.

[18] Suzuki K. and Nakazawa M. (2001). Ultrabroad band white light generation from a multimode photonic bandgap fiber with an air core. *Proc. Conf. Laser and Electro-Optics* (CLEO) Pasific Rim, WIPD1-11.

[19] Suzuki K., Kubota H., Kawanishi S., Tanaka M., and Fujita M. (2001). High-speed bi-directional polarisation division multiplexed optical transmission in ultra low-loss (1.3 dB/km) polarisation-maintaining photonic crystal fibre. *Electron. Lett.*, 37 (23), 1399-1400.

[20] Lizier J. T. and Town G. E. (2001). Splice losses in holey optical fibers. *IEEE Photon. Technol. Lett.*, 13 (8), 794-796.

[21] Benabid F., Knight J. C., Antonopoulos G., and Russell P. St. J. (2002). Stimulated Raman scattering in hydrogen-filled hollow-core photonic crystal fiber. *Science*, 298, 399-402.

[22] Temelkuran B., Hart S. D., Benoit G., Joannopoulos J. D., and Fink Y. (2002). Wavelength-scalable hollow optical fibres with largw photonic bandgaps for CO_2 laser transmission. *Nature*, 420, 650-653.

[23] Litchinitser N. M., Abeeluck A. M., Headley C., and Eggleton B. J. (2002). Antiresonant reflecting photonic crystal optical waveguides. *Opt. Lett.*, 27 (18), 1592-1594.

[24] Reeves W. H., Knight J. C., Russell P. St. J., and Roberts P. J. (2002). Demonstration of ultra-flattened dispersion in photonic crystal fibers. *Opt. Express*, 10 (14), 609-613.

[25] Price J. H., Furusawa K., Monro T. M., Lefort L., and Richardson D. J. (2002). Tunable, femtosecond pulse source operating in the range 1.06-1.33 μm based on an Yb^{3+}-doped holey fiber amplifier. *J. Opt. Soc. Am. B*, 19 (6), 1286-1294.

[26] Smith C. M., Venkataraman N., Gallagher M. T., Müller D., West J. A., Borrelli N. F., Allan D. D., and Koch K. W. (2003). Low-loss hollow-core silica/air photonic bandgap fibre. *Nature*, 424, 657-659.

[27] Russell P. St. J. (2003). Photonic crystal fibers. *Science*, 299, 358-362.

[28] Knight J. C. (2003). Photonic crystal fibres. *Nature*, 424, 847-851.

[29] Ouzounov D. G., Ahmad F. R., Müller D., Venkataraman N., Gallagher M. T., Thomas M. G., Silcox J., Koch K. W., and Gaeta A. L. (2003). Generation of megawatt optical solitons in hollow-core photonic band-gap fibers. *Science*, 301, 1702-1704.

[30] Limpert J., Schreiber T., Nolte S., Zellmer H., Tünnerman A., Iliew R., Lederer F., Broeng J., Vienne G., Petersson A., and Jakobsen C. (2003). High-power air-clad large-mode-area photonic crystal fiber laser. *Opt. Express*, 11 (7), 818-823.

[31] Saitoh K., Koshiba M., Hasegawa T., and Sasaoka E. (2003). Chromatic dispersion control in photonic crystal fibers: application to ultra-flattened dispersion. *Opt. Express*, 11 (8), 843-852.

[32] Hansen K. P. (2003). Dispersion flattened hybrid-core nonlinear photonic crystal fiber. *Opt. Express*, 11 (13), 1503-1509.

[33] Tajima K., Zhou J., Kurokawa K., and Nakajima K. (2003). Low water peak photonic crystal fibers. *Proc. Eur. Conf. Opt. Commun.* (ECOC2003), Th4.1.6.

[34] Zsigri B., Peucheret C., Nielsen M.~D., and Jeppesen P. (2003). Transmission over 5.6 km large effective area and low-loss (1.7 dB/km) photonic crystal fibre. *Electron. Lett.*, 39 (10), 796-798.

[35] Peucheret C., Zsigri B., Andersen P. A., Berg K. S., Tersigni A., Jeppesen P., Hansen K. P., and Nielsen M. D. (2003). 40 Gbit/s transmission over photonic crystal fibre using mid-span spectral inversion in highly nonlinear photonic crystal fibre. *Electron. Lett.*, 39 (12), 919-921.

[36] Limpert J., Liem A., Reich M., Schreiber T., Nolte S., Zellmer H., Tünnerman A., Broeng J., Petersson A., and Jakobsen C. (2004). Low-nonlinearity single-transverse-mode ytterbium-doped photonic crystal fiber amplifier. *Opt. Express,* 12 (7), 1313-1319.

[37] Nielsen M. D., Jacobsen C., Mortensen N.~A., Folkenberg J. R., and Simonsen H. R. (2004). Low-loss photonic crystal fibers for transmission systems and their dispersion properties. *Opt. Express*, 12 (7), 1372-1376.

[38] Humbert G., Knight J. C., Bouwmans G., Russell P. St. J., Williams D. P., Roberts P. J., and Mangan B. J. (2004). Hollow core photonic crystal fibers for beam delivery. *Opt. Express*, 12 (8), 1477-1484.

[39] Vienne G., Xu Y., Jackobsen C., Deyerl H., Jensen J. B., Sorensen T., Hansen T. P., Huang Y., Terrel M., Lee R. K., Mortensen N. A., Broeng J., Simonsen H., Bjarkey A., and Yariv A. (2004). Ultra-large bandwidth hollow-core guiding in all-silica Bragg fibers with nano-supports. *Opt. Express*, 12 (15), 3500-3508.

[40] Tajima K., Zhou J., Nakajima K., and Sato K. (2004). Ultralow loss and long length photonic crystal fiber. *IEEE J. Lightwave Technol.*, 22 (1), 7-10.

[41] Kim H. K., Shin J., Fan S., Digonnet M. J. F., and Kino G. S. (2004). Designing air-core photonic-bandgap fibers free of surface modes. *IEEE J. Quantum Electron.*, 40 (5), 551--556.

[42] Roberts P. J., Couny F., Sabert H., Mangan B. J., Williams D. P., Farr L., Mason M. W., Tomlinson A., Birks T. A., Knight J. C., and Russell P. St. J. (2005). Ultimate low loss of hollow-core photonic crystal fibers. *Opt. Express*, 13 (1), 236-244.

[43] Poletti F., Broderick G. R., and Richardson D. J. (2005). The effect of core asymmetries on the polarization properties of hollow core photonic bandgap fibers. *Opt. Express*, 13 (22), 9116-9124.

[44] Nakajima K., Zhou J., Tajima K., Kurokawa K., Fukai C., and Sankawa I. (2005). Ultrawide-band single-mode transmission performance in a low-loss photonic crystal fiber. *IEEE J. Lightwave Technol.*, 23 (1), 7-12.

[45] Zhou J., Tajima K., Nakajima K., Kurokawa K., Fukai C., Matsui T., and Sankawa I. (2005). Progress on low loss photonic crystal fibers. *Opt. Fiber Technol.*, 11, 101-110.

[46] Kurokawa K., Tajima K., Tsujikawa K., Nakajima K., Matsui T., Sankawa I., and Haibara T. (2006). Penalty-free dispersion-managed soliton transmission over a 100-km low-loss PCF. *IEEE J. Lightwave Technol.*, 24 (1), 32-37.

[47] Russell P. St. J. (2006). Photonic-crystal fibers. *IEEE J. Lightwave Technol.*, 24 (12), 4729-4749.

[48] Benabid F. (2006). Hollow-core photonic bandgap fibre: new light guidance for new science and technology. *Philos. Trans. R. Soc. A*, 364, 3439-3462.

[49] Couny F., Benabid F., and Light P. S. (2006). Large-pitch kagome-structured hollow-core photonic crystal fiber. *Opt. Lett.*, 31 (24), 3574-3576.

[50] Dudley J. M., Genty G., and Coen S. (2006). Supercontinuum generation in photonic crystal fiber. *Rev. Mod. Phys.*, 78 (4), 1135-1184.

[51] Hasegawa H., Oikawa Y., Yoshida M., Hirooka T., and Nakazawa M. (2006). 10 Gb/s transmission over 5 km at 850 nm using single-mode photonic crystal fiber, single-mode VCSEL, and Si-APD. IEICE Electron. *Express*, 3 (6), 109-114.

[52] Knight J. C. (2007). Photonic crystal fiber and fiber lasers. *J. Opt. Soc. Am. B*, 24 (8), 1661-1668.

[53] Couny F., Benabid F., Roberts P. J., Light P. S., and Raymer M. G. (2007). Generation and photonic guidance of multi-octave optical-frequency combs. *Science*, 318, 1118-1121.

[54] Hensley C. J., Ouzounov D. G., Gaeta A. L., Venkataraman N., Gallagher M. T., and Koch K. W. (2007). Silica-glass contribution to the effective nonlinearity of hollow-core photonic band-gap fibers. *Opt. Express*, 15 (6), 3507-3512.

[55] Argyros A. and Pla J. (2007). Hollow-core polymer fibres with a kagome lattice: potential for transmission in the infrared. *Opt. Express*, 15 (12), 7713-7719.

[56] Nakajima K., Matsui T., Kurokawa K., Tajima K., and Sankawa I. (2007). High-speed and wideband transmission using dispersion-compensating/managing photonic crystal fiber and dispersion-shifted fiber. *IEEE J. Lightwave Technol.*, 25 (9), 2719-2726.

[57] Oikawa Y., Hasegawa H., Suzuki K., Inoue Y., Hirooka T., and Nakazawa M. (2007). 4 × 10 Gb/s WDM transmission over a 5-km-long photonic crystal fiber in the 800-nm region. *IEEE Photon. Technol. Lett.*, 17 (8), 613-615.

[58] Ieda K., Kurokawa K., Tajima K., and Nakajima K. (2007). Visible to infrared high-speed WDM transmission over PCF. *IEICE Electron. Express*, 4 (12), 375-379.

[59] Kurokawa K., Tsujikawa K., Tajima K., Nakajima K., and Sankawa I. (2007). 10 Gb/s WDM transmission at 1064 and 1550 nm over 24 km photonic crystal fiber with negative power penalties. *IEICE Trans. Commun.*, E90-B (10), 2803-2808.

[60] Kurokawa K., Nakajima K., Tsujikawa K., Yamamoto T., and Tajima K. (2009). Ultra-wideband transmission over low loss PCF. *IEEE J. Lightwave Technol.*, 27 (11), 1635-1662.

[61] Villatoro J., Kreuzer M. P., Jha R., Minkovich V. P., Finazzi V., Badenes G., and Prunert V. (2009). Photonic crystal fiber interferometer for chemical vapor detection with high sensitivity. *Opt. Express*, 17 (3), 1447-1453.

[62] Wang Y. Y., Wheeler N. V., Couny F., Roberts P. J., and Benabid F. (2011). Low loss broadband transmission in hypocycloid-core kagome hollow-core photonic crystal fiber. *Opt. Lett.*, 36 (5), 669-671.

[63] Pryamikov A. D., Biriukov A. S., Kosolapov A. F., Plotnichenko V. G., Semjonov S. L., and Dianov E. M. (2011). Demonstration of a waveguide regime for a silica hollow-core microstructured optical fiber with a negative curvature of the core boundary in the spectral region > 3.5 □m. *Opt. Express*, 19 (2), 1441-1448.

[64] Kosolapov A. F., Pryamikov A. D., Biriukov A. S., Shiryaev V. S., Astapovich M. S., Snopatin G. E., Plotnichenko V. G., Churbanov M. F., and Dianov E. M. (2011). Demonstration of CO_2-laser power delivery through chalcogenide-glass fiber with negative-curvature hollow core. *Opt. Express*, 19 (25), 25723-25728.

[65] Benabid F. and Roberts P. J. (2011). Linear and nonlinear optical properties of hollow core photonic crystal fiber. *J. Mod. Opt.*, 58 (2), 87-124.

[66] Wang Y. Y., Peng X., Alharbi M., Fourcade-Dutin C., Bradley T. D., Gérôme F., Mielke M., Booth T., and Benabid F. (2012). Design and fabrication of hollow-core photonic crystal fibers for high-power ultrashort pulse transportation and pulse compression. *Opt. Lett.*, 37 (15), 3111-3113.

[67] Bradley T. D., Wang Y., Alharbi M., Debord B., Fourcade-Dutin C., Beaudou B., Gerome F., and Benabid F. (2013). Optical properties of low loss (70dB/km) hypocycloid-core kagome hollow core photonic crystal fiber for Rb and Cs based optical applications. *IEEE J. Lightwave Technol.*, 31 (16), 2752-2755.

[68] Frosz M. H., Nold J., Weiss T., Stefani A., Babic F., Rammler S., and Russell P. St. J. (2013). Five-ring hollow-core photonic crystal fiber with 1.8 dB/km loss. *Opt. Lett.*, 38 (13), 2215-2217.

[69] Kolyadin A. N., Kosolapov A. F., Prymikov A. D., Biriukov A. S., Plotnichenko V. G., and Dianov E. M. (2013). Light transmission in negative curvature hollow core fiber in extremely high material loss region. *Opt. Express*, 21 (8), 9514-9519.

[70] Yu F. and Knight J. C. (2013). Spectral attenuation limits of silica hollow core negative curvature fiber. *Opt. Express*, 21 (18), 21466-21471.

[71] Poletti F., Wheeler N. V., Petrovich M. N., Baddela N., Fokoua E. N., Hayes J. R., Gray D. R., Li Z., Slavik R., and Richardson D. J. (2013). Towards high-capacity fibre-optic communications at the speed of light in vacuum. *Nature Photon.*, 7, 279-284.

[72] Belardi W. and Knight J. C. (2014). Hollow antiresonant fibers with reduced attenuation. *Opt. Lett.*, 30 (7), 1853-1856.

[73] Belardi W. and Knight J. C. (2014). Hollow antiresonant fibers with low bending loss. *Opt. Express*, 22 (8), 10091-10096.

[74] Debord B., Alharbi M., Vincetti L., Husakou A., Fourcade-Dutin C., Hoenninger C., Mottay E., Gérôme F., and Benabid F. (2014). Multi-meter fiber-delivery and pulse self-compression of milli-Joule femtosecond laser and fiber-aided laser-micromachining. *Opt. Express*, 22 (9), 10735-10746.

[75] Poletti F. (2014). Nested antiresonant nodeless hollow core fiber. *Opt. Express,* 22 (20), 23807-23828.

[76] Hartung A., Kobelke J., Schwuchow A., Wondraczek K., Bierlich J., Popp J., Frosch T., and Schmidt M. A. (2014). Double antiresonant hollow core fiber-guidance in the deep ultraviolet by modified tunneling leaky modes. *Opt. Express,* 22 (16), 19131-19140.

[77] Belardi W. (2015). Design and properties of hollow antiresonant fibers for the visible and near infrared spectral range. *IEEE J. Lightwave Technol.*, 33 (21), 4497-4503.

[78] Balciunas T., Fourcade-Dutin C., Fan G., Witting T., Voronin A. A., Zheltikov A. M., Gerome F., Paulus G. G., Baltuska A., and Benabid F. (2015). A strong-field driver in the single-cycle regime based on self-compression in a kagome fibre. *Nature Commun.*, 6, 6117-1-6117-7.

[79] Jiang X., Joly N. Y., Finger M. A., Babic F., Wong G. K. L., Travers J. C., and Russell P. St. J. (2015). Deep-ultraviolet to mid-infrared supercontinuum generated in solid-core ZBLAN photonic crystal fibre. *Nature Photon.,* 9, 133-139.

[80] Kosolapov A. F., Alagashev G. K., Kolyadin A. N., Pryamikov A. D., Biryukov A. S., Bufetov I. A., and Dianov E. M. (2016). Hollow-core revolver fibre with a double-capillary reflective cladding. *Quantum Electron.*, 46 (3), 267-270.

[81] Yu F. and Knight J. C. (2016). Negative curvature hollow-core optical fiber. *IEEE J. Selected Top. Quantum Electron.*, 22 (2), 4400610.

[82] Yu F., Xu M., and Knight J. C. (2016). Experimental study of low-loss single-mode performance in anti-resonant hollow-core fibers. *Opt. Express*, 24 (12), 12969-12975.

[83] Debord B., Amsanpally A., Chafer M., Baz A., Maurel M., Blondy J. M., Hugonnot E., Scol F., Vincetti L., Gérôme F., and Benabid F. (2017). Ultralow transmission loss in inhibited-coupling guiding hollow fibers. *Optica*, 4 (2), 209-217.

[84] Wheeler N. V., Bradley T. D., Hayes J. R., Gouveia M. A., Liang S., Chen Y., Sandoghchi S. R., Mousavi S. M. A., Poletti F., Petrovich M. N., and Richardson D. J. (2017). Low-loss kagome hollow-core fibers operating from the near- to the mid-IR. *Opt. Lett.*, 42 (13), 2571-2574.

[85] Wei C., Weiblen R. J., Menyuk C. R., and Hu J. (2017). Negative curvature fibers. *Ad. Opt. Photon.*, 9 (3), 504-561.

[86] Edavalath N. N., Gunendi M. C., Beravat R., Wong G. K. L., Frosz M. H., Menard J. M., and Russell P. St. J. (2017). Higher-order mode suppression in twisted single-ring hollow-core photonic crystal fibers. *Opt. Lett.*, 42 (11), 2074-2077.

[87] Maurel M., Chafer M., Amsanpally A., Adnan M., Amrani F., Debord B., Vincetti L., Gérôme F., and Benabid F. (2018). Optimized inhibited-coupling kagome fibers at Yb-Nd:Yag (8.5 dB/km) and Ti:Sa (30 dB/km) ranges. *Opt. Lett.*, 43 (7), 1598-1601.

[88] Bradley T. D., Hayes J. R., Chen Y., Jasion G. T., Sandoghchi S. R., Slavik R., Fokoua E. N., Bawn S., Sakr H., Davidson I. A., Taranta A., Thomas J. P., Petrovich M. N., Richardson D. J., and Poletti F. (2018). Record low-loss 1.3 dB/km data transmitting antiresonant hollow core fibre. *Proc. European Conf. Opt. Commun.* (ECOC 2018), 1-3.

[89] Bufetov I. A., Kosolapov A. F., Pryamikov A. D., Gladyshev A. V., Kolyadin A. N., Krylov A. A., Yatsenko Y. P., and Biriukov A. S. (2018). Revolver hollow core optical fibers. *Fibers*, 6, 39-1-39-26.

[90] Bradley T., Hayes J., Hooper L., Sakr H., Jasion G., Alonso M., Taranta A., Saljoghei A., Fake M., Davidson I., Chen Y., Wheeler N., Fokoua E., Wang W., Sandoghchi S. R., Richardson D., Poletti F., and Mulvad H. C. (2019). Hollow core fibre with 0.65 dB/km attenuation across the C and L telecommunication bands. *Proc. European Conf. Opt. Commun.* (ECOC 2019), PD.3.1.

[91] Yu F., Song P., Wu D., Birks T., Bird D., and Knight J. C. (2019). Attenuation limit of silica-based hollow-core fiber at mid-IR wavelengths. *Appl. Phys. Lett. Photon.*, 4, 080803-1-080803-6.

[92] Debord B., Amrani F., Vincetti L., Gérôme F., and Benabid F. (2019). Hollow-core fiber technology: the rising of 'gas photonics.' *Fibers*, 7, 16-1-16-58.

[93] Sopalla R. P., Wong G. K. L., Joly N. Y., Frosz M. H., Jiang X., Ahmed G., and Russell P. St. J. (2019). Generation of broadband circularly polarized supercontinuum light in twisted photonic crystal fibers. *Opt. Lett.*, 44 (16), 3964-3967.

[94] Abdolghader P., Pegoraro A. F., Joly N. Y., Ridsdale A., Lausten R., Legare F., and Stolow A. (2020). All normal dispersion nonlinear fibre supercontinuum source characterization and application in hyperspectral stimulated Raman scattering microscopy. *Opt. Express*, 28 (24), 35997-36008.

[95] Sakr H., Chen Y., Jasion G. T., Bradley T. D., Hayes J. R., Mulvad H. C. H., Davidson I. A., Fokoua E. N., and Poletti F. (2020). Hollow core optical fibres with

comparable attenuation to silica fibres between 600 and 1100 nm. *Nat. Commun.*, 11, 6030.

[96] Jasion G. T., Bradley T. D., Harrington K., Sakr H., Chen Y., Fokoua E. N., Davidson I. A., Taranta A., Hayes J. R., Richardson D. J., and Poletti F. (2020). Hollow core NANF with 0.28 dB/km attenuation in the C and L bands. *Proc. Opt. Fiber Conf.* (OFC 2020), Th4B.4.

[97] Koeppel M., Sharma A., Podschus J., Sundaramahalingam S., Joly N. Y., Xie S., Russell P. St. J., and Schmauss B. (2021). Doppler optical frequency domain reflectometry for remote fiber sensing. *Opt. Express*, 29 (10), 14615-14629.

[98] Amrani F., Osono J. H., Delahaye F., Giovanardi F., Vincetti L., Debord B., Gérôme F., and Benabid F. (2021). Low-loss single-mode hybrid-lattice hollow-core photonic-crystal fibre. *Light Sci. Applications*, 10, 7-1-7-12.

[99] Debord B., Alharbi M., Bradley T. D., Fourcade-Dutin C., Wang Y. Y., Vincetti L., Gérôme F., and Benabid F. (2013). Hypocycloid-shaped hollow-core photonic crystal fiber Part I: arc curvature effect on confinement loss. *Opt. Express*, 21 (23), 28597-28608.

[100] Wang Y. and Ding W. (2017). Confinement loss in hollow-core negative curvature fiber: A multi-layered model. *Opt. Express*, 25 (26), 33122-33133.

[101] Yu R., Chen Y., Shui L., and Xiao L. (2020). Hollow-core photonic crystal fibers gas sensing. *Sensors*, 20, 2996.

[102] Nampoothiri A. V. V., Jones A. M., Fourcade-Dutin C., Mao C., Dadashzadeh N., Baumgart B., Wang Y. Y., Alharbi M., Bradley T., Campbell N., Benabid F., Washburn B. R., Corwin K. L., and Rudolph W. (2012). Hollow-core optical fiber gas lasers (HOFGLAS): a review. *Opt. Mat. Express*, 2 (7), 948-961.

[103] Benoit A., Ilinova E., Beaudou B., Debord B., Gerôme F., and Benabid F. (2017). Spectral-temporal dynamics of high power Raman picosecond pulse using H_2-filled kagome HC-PCF. *Opt. Lett.*, 42 (19), 3896-3899.

[104] Cao L., Gao S., Peng Z., Wang X., Wang Y., and Wang P. (2018). High peak power 2.8 μm Raman laser in a methane-filled negative-curvature fiber. *Opt. Express*, 26 (5), 5609-5615.

[105] Li Z., Huang W., Cui Y., and Wang Z. (2018). Efficient mid-infrared cascade Raman source in methane-filled hollow-core fibers operating at 2.8 μm. *Opt. Lett.*, 43 (19), 4671-4674.

[106] Gladyshev A. V., Kosolapov A. F., Khudyakov M. M., Yatsenko Y. P., Kolyadin A. N., Krylov A. A., Pryamikov A. D., Biriukov A. S., Likhachev M. E., Bufetov I. A., and Dianov E. M. (2018). 2.9, 3.3, and 3.5 μm Raman lasers based on revolver hollow-core silica fiber filled by $^1H_2/D_2$ gas mixture. *IEEE J. Selected Topics Quantum Electron.*, 24 (3), 0903008.

[107] Astapovich M. S., Gladyshev A. V., Khudyakov M. M., Kosolapov A. F., Likhachev M. E., and Bufetov I. A. (2019). Watt-level nanosecond 4.42-μm Raman laser based on silica fiber. *IEEE Photon. Technol. Lett.*, 31 (1), 78-81.

[108] Wang Y., Dasa M. K., Adamu A. I., Antonio-Lopez J. E., Habib M. S., Amezcua-Correa R., Bang O., and Markos C. (2020). High pulse energy and quantum efficiency mid-infrared gas Raman fiber laser targeting CO_2 absorption at 4.2 μm. *Opt. Lett.*, 45 (7), 1938-1941.

[109] Gladyshev A., Yatsenko Y., Kolyadin A., Kompanets V., and Bufetov I. (2020). Mid-infrared 10-μJ-level sub-picosecond pulse generation via stimulated Raman scattering in a gas-filled revolver fiber. *Opt. Mat. Express*, 10 (12), 3081-3089.

[110] Beaudou B., Gerôme F., Wang Y. Y., Alharbi M., Bradley T. D., Humbert G., Auguste J.-L., Blondy J.-M., and Benabid F. (2012). Millijoule laser pulse delivery for spark ignition through kagome hollow-core fiber. *Opt. Lett.*, 37 (9), 1430-1432.

[111] Jaworski P., Yu F., Maier R. R. J., Wadsworth W. J., Knight J. C., Shephard J. D., and Hand D. P. J. (2013). Picosecond and nanosecond pulse delivery through a hollow-core negative curvature fiber for micro-machining applications. *Opt. Express*, 21 (19), 22742-22753.

[112] Ulrich A., Maier R. R. J., Yu F., Knight J. C., Hand D. P. J., and Shephard J. D. (2013). Flexible delivery of Er:YAG radiation at 2.94 μm with negative curvature silica glass fibers: a new solution for minimally invasive surgical procedures. *Biomedical Opt. Express*, 4 (2), 193-205.

[113] Dumitrache C., Rath J., and Yalin A. P. (2014). High power spark delivery system using hollow core kagome lattice fibers. *Materials*, 7, 5700-5710.

[114] Michieletto M., Lyngso J. K., Jakobsen C., Laegsgaard J., Bang O., and Alkeskjold T. T. (2016). Hollow-core fibers for high power pulse delivery. *Opt. Express*, 24 (7), 7103-7119.

[115] Elizer S. and Wedel B. (2018). Hollow core optical fibers for industrial ultra short pulse laser beam delivery applications. *Fibers*, 6, 80-1-80-11.

[116] Zhu X., Wu D., Wang Y., Yu F., Li Q., Qi Y., Knight J., Chen S., and Hu L. (2021). Delivery of CW laser power up to 300 Watts at 1080 nm by an uncooled low-loss ant-resonant hollow-core fiber. *Opt. Express*, 29 (2), 1492-1501.

[117] Kashyap R. and Blow K. J. (1988). Observation of catastrophic self-propelled self-focusing in optical fibres. *Electron. Lett.*, 24 (1), 47-49.

[118] Kashyap R. (1988). Self-propelled self-focusing damage in optical fibres. *Lasers'87: Proc. Xth Int. Conf. Lasers and Applications*, 859-866, STS Press, McLean.

[119] Hand D. P. and Russell P. St. J. (1988). Solitary thermal shock waves and optical damage in optical fibers: the fiber fuse. *Opt. Lett.*, 13 (9), 767-769.

[120] Hand D. P., Townsend J. E., and Russell P. St. J. (1988). Optical damage in fibers: the fiber fuse. *Proc. Conf. on Lasers and Electro-Optics* (CLEO), WJ1.

[121] Hand D. P. and Russell P. St. J. (1988). Soliton-like thermal shock-waves in optical fibres: origin of periodic damage tracks. *Proc. Eur. Conf. Opt. Commun.*, 111-114.

[122] Kolyadin A. N., Kosolapov A. F., and Bufetov I. A. (2018). Optical discharge propagation along hollow-core optical fibres. *Quantum Electron.*, 48 (12), 1138-1142.

[123] Bufetov I. A., Kolyadin A. N., Kosolapov A. F., Efremov V. P., and Fortov V. E. (2019). Catastrophic damage in hollow core optical fibers under high power laser radiation. *Opt. Express*, 27 (13), 18296-18311.

[124] Shuto Y. (2015). Simulation of fiber fuse phenomenon in single-mode optical fibers. In *Advances in Optical Fiber Technology*, Eds. Yasin M., Arof H., and Harun S. W. Chap. 5, InTech, Rijeka.

[125] Thiyagarajan M. and Thompson S. (2012). Optical breakdown threshold investigation of 1064 nm laser induced air plasma. *J. Appl. Phys.*, 111 (7), 073302-1-073302-8.

[126] Sticker J. and Parker J. G. (1982). Experimental investigation of electrical breakdown in nitrogen and oxygen induced by focused laser radiation at 1.064 μ. *J. Appl. Phys.*, 53 (2), 851-855.

[127] Giuliano C. R. (1964). Laser-induced damage to transparent dielectric materials. *Appl. Phys. Lett.*, 5 (7), 137-139.

[128] Fersman I. A. and Khazov K. D. (1970). Damage of transparent dielectric surfaces by a laser beam. *Sov. Phys. Tech. Phys.*, 15, 834-838.

[129] Giuliano C. R. (1972). Laser-induced damage in transparent dielectrics: The relationship between surface damage and surface plasmas. *IEEE J. Quantum Electron.*, QE-8 (9), 749-754.

[130] Crisp M. D., Boling N. L., and Dubé G. (1972). Importance of Fresnel reflections in laser surface damage of transparent dielectrics. *Appl. Phys. Lett.*, 22 (8), 364-366.

[131] Boling N. L., Crisp M. D., and Dubé G. (1973). Laser induced surface damage. *Appl. Opt.*, 12 (4), 650-660.

[132] Bloembergen N. (1974). Laser-induced electric breakdown in solids. *IEEE J. Quantum Electron.*, QE-10 (3), 375-386.

[133] Bloembergen N. (1973). Role of cracks, pores, and absorbing inclusions on laser induced damage threshold at surfaces of transparent dielectrics. *Appl. Opt.*, 12 (4), 661-664.

[134] Génin F. Y., Salleo A., Pistor T. V., and Chase L. L. (2001). Role of light intensification by cracks in optical breakdown on surfaces. *J. Opt. Soc. Am. A,* 18 (10), 2607-2616.

[135] Smith A. V. and Do B. T. (2008). Bulk and surface laser damage of silica by picosecond and nanosecond pulses at 1064 nm. *Appl. Opt.*, 47 (26), 4812-4832.

[136] Kaiser P., Tynes A. R., Astle H. W., Pearson A. D., French W. G., Jaeger R. E., and Cherin A. H. (1973). Spectral losses of unclad vitreous silica and soda-lime-silicate fibers. *J. Opt. Soc. Am.*, 63 (9), 1141-1148.

[137] Schultz P. C. (1974). Optical absorption of the transition elements in vitreous silica. *J. Am. Ceram. Soc.*, 57 (7), 309-313.

[138] Hopper R. W. and Uhlmann D. R. (1970). Mechanism of inclusion damage in laser glass. *J. Appl. Phys.*, 41 (10), 4023-4037.

[139] Shoji M. (1995). *Heat Transfer Textbook*. Appendix F, University of Tokyo Press, Tokyo.

[140] Schick H. L. (1960). A thermodynamic analysis of the high-temperature vaporization properties of silica. *Chem. Rev.*, 60, 331-362.

[141] Hidai H., Wada J., Iwamoto T., Matsusaka S., Chiba A., Kishi T., and Morita N. (2016). Experimental and theoretical study on the driving force and glass flow by laser-induced metal sphere migration in glass. *Sci. Rep.*, 6, 38545.

[142] Davis D. D., Mettler S. C., and DiGiovani D. J. (1996). A comparative evaluation of fiber fuse models. *Proc. Soc. Photo-Opt. Instrum. Eng.*, 2966, 592-606.

[143] Jensen B. (1985). The quantum extension of the Drude-Zener theory in polar semiconductors. In *Handbook of Optical Constants of Solids*, Ed. Palik E. D., Chap. 9. Academic Press Inc., New York.

[144] Ordal M. A., Long L. L., Bell R. J., Bell S. E., Bell R. R., Alexander, Jr. R. W., and Ward C. A. (1983). Optical properties of metals Al, Co, Cu, Au, Fe, Pb, Ni, Pd, Pt, Ag, Ti, and W in the infrared and far infrared. *Appl. Opt.*, 22 (7), 1099-1119.

[145] Shuto Y. (2016). End face damage and fiber fuse phenomena in single-mode fiber-optic connectors. *J. Photonics*, 2016, 2781392.

[146] Sakai J. and Suzuki Y. (2011). Equivalence between in-phase and antiresonant reflection conditions in Bragg fiber and its application to antiresonant reflecting optical waveguide-type fibers. *J. Opt. Soc. Am. B,* 28 (1), 183-192.

[147] Argyros A. (2002). Guided modes and loss in Bragg fibres. *Opt. Express*, 10 (24), 1411-1417.

[148] Deng A., Hasan M. I., Wang Y., and Chang W. (2020). Analyzing mode index mismatch and field overlap for light guidance in negative-curvature fibres. *Opt. Express*, 28 (19), 27974-27988.

[149] Shibata N. and Edahiro T. (1982). Refractive-index dispersion for GeO_2-, P_2O_5- and B_2O_3-doped silica glasses in optical fibers. *Trans. IECE Jpn.*, E65 (3), 166-172.

[150] Atkins R. M., Simpkins P. G., and Yabon A. D. (2003). Track of a fiber fuse: a Rayleigh instability in optical waveguides. *Opt. Lett.*, 28 (12), 974-976.

[151] Todoroki S. (2014). Fiber Fuse: Light-Induced Continuous Breakdown of Silica Glass Optical Fiber. Chap. 3, *NIMS Monographs*, Springer, Tokyo.

[152] Ramsden S. A. and Savic P. (1964). A radiative detonation model for the development of a laser-induced spark in air. *Nature*, 203 (4951), 1217-1219.

[153] Sedov L. I. (1959). *Similarity and Dimensional Methods in Mechanics*. Chap. 4, Academic Press, New York.

[154] Incropera F. P. and DeWitt D. P. (2002). *Introduction to Heat Transfer*. 4th Ed. Chap. 5, John Wiley & Sons, Inc., New York.

[155] Shoji M. (1995). *Heat Transfer Textbook*. Chap. 4, University of Tokyo Press, Tokyo.

Index

A

B

C

D

F

G

O

P

R

S

T

U

V

W